The Theoretical Individual

Michael Charles Tobias • Jane Gray Morrison

The Theoretical Individual

Imagination, Ethics and the Future
of Humanity

Michael Charles Tobias
Dancing Star Foundation
Los Angeles, CA, USA

Jane Gray Morrison
Dancing Star Foundation
Los Angeles, CA, USA

Please note that copyright will be vested in name of Authors as specified in publishing contract
ISBN 978-3-319-71442-4   ISBN 978-3-319-71443-1 (eBook)
https://doi.org/10.1007/978-3-319-71443-1

Library of Congress Control Number: 2017960837

© Michael Charles Tobias and Jane Gray Morrison 2018
This work is subject to copyright. All rights are reserved by the Publisher, whether the whole or part of the material is concerned, specifically the rights of translation, reprinting, reuse of illustrations, recitation, broadcasting, reproduction on microfilms or in any other physical way, and transmission or information storage and retrieval, electronic adaptation, computer software, or by similar or dissimilar methodology now known or hereafter developed.
The use of general descriptive names, registered names, trademarks, service marks, etc. in this publication does not imply, even in the absence of a specific statement, that such names are exempt from the relevant protective laws and regulations and therefore free for general use.
The publisher, the authors and the editors are safe to assume that the advice and information in this book are believed to be true and accurate at the date of publication. Neither the publisher nor the authors or the editors give a warranty, express or implied, with respect to the material contained herein or for any errors or omissions that may have been made. The publisher remains neutral with regard to jurisdictional claims in published maps and institutional affiliations.

Jesse Nusbaum, "Santa Fe Plaza in the Winter" circa 1912, Courtesy Palace of the Governors Photo Archives (NMHM/DCA), #061456

Printed on acid-free paper

This Springer imprint is published by Springer Nature
The registered company is Springer International Publishing AG
The registered company address is: Gewerbestrasse 11, 6330 Cham, Switzerland

Preface

Abstract The basic narrative and seminal questions of the book are first posed. What comprises an individual, particularly a human individual? To what extent has human history provided empirical evidence for the capacity of an individual to exert meaningful suasion over her/his species? If so, by what means? If not, what then?

Prognostications

Let us imagine that every conceivable circumstance pertaining to natural history has confronted *Homo sapiens* during their 300,000 year + regime (acknowledging the 22 human fossils from Jebel Irhoud in western Morocco). The durability of their questions, answers, and choices has been furthered along by genetic deep lineages dating back as far as one is likely to concede, to the origins of single and then multicellular life, the evolution of neurons, the development of an assured infinitude of communication pathways, and other life-fostering orientations.

Quintessentially, those sensitive spots that collectively make for the great amphitheater of the biosphere have given us surpassing awe, suffering, and joy, as well as the fair warnings that have punctuated our tumultuous journey.

It turns out that those sensitive spots – in human beings – have too easily transmogrified into catastrophic blind spots.

In this epic saga sweeping a small planet, nameless in almost every respect, we (our species) wish for hallowed recognition amid the biological topsy-turvy of as many legacies as there are ideas, convictions, and ideals, as multiplied by the trillions of individual organisms that have enshrined, harbored, and nurtured everything that we know to be, out of living vapors. The resulting task is akin to an ether of imagination stirred toward survival and incessantly rephrased in the guise of a profound and uneasy question.

It has been our fate to differentiate betwixt that steady proliferation of life forms, with all the consciousness our demonstrable exercise of reflexes has been capable of. Because we have no measurements or even baseline for consciousness, as such, our sudden journey does not comprise great thought, just thought, neither consistent virtue nor villainy, just a multitude of behaviors. We heed whatever compass reading is convenient, restlessly grappling with those semblances of order and invention our myriad compulsions have seized upon, from day to day, millennium by millennium. During the last decade of Charles Darwin's life, his two most important books were *The Expression of the Emotions in Man and Animals* (John Murray, London, 1872) and *The Formation of Vegetable Mould Through the Action of Worms, with Observations on Their Habits* (John Murray, London, 1881). He had studied these topics for most of his life and was clear about a crucial insight: all other species contain brilliant and individualist lives, in the same general manner as us humans. Such thinking contradicted most of Western science, which was – when it came to the topic of plants and animals and their minds – of a fully Cartesian orientation, the steadfast refusal to grant self-aware individualism to any organism other than human ones. One year prior to Darwin's publication of *The Variation of Animals and Plants Under Domestication* (January 1868), Charles Henry Turner was born in Cincinnati, Ohio. The first African-American to obtain a Ph.D. in zoology from the University of Chicago, he was the first person to recognize and prove that insects hear and learn, altering future behavior upon learned experience, and that honey bees see color. This great scientist will be remembered for, among many other things, his humility before and adoration for the largest number of creatures on the face of the planet. He ascribed individualism to each of them. The generous attribution of such limitless sentience will form a crucial pillar of this work.[1] Turner was a great contemporary of the Carlux (Perigord/Dordogne, France) zoologist Pierre-Paul Grassé whose assiduous 52-volume *Traité de Zoologie* demonstrated an undying love of termites, the most profound socialists on the planet, whose typically oppositional orientation to territory, other than that of ants has become the backdrop for the most pronounced scientific perturbations ever advanced. E. O. Wilson has made much of the battlegrounds between these two multitudinous groups of insects. Albrecht Altdorfer painted the human equivalent in his "The Battle of Alexander at Issus" (1529, Alte Pinakothek, Munich) in which historians have adduced over 125,000 combatants illustrated in the glorious, if blood-swathed, work.

Equally impressive, if not more so than Darwin's and Turner's interpolations, were the fully documented acuities of Pennsylvania-born Theodore Albert Parker III (1953–1993), an ornithologist/sage who could identify over 4000 bird species solely by their songs and calls. He has been described as "the greatest specialist on the life histories of neotropical birds there ever was."[2]

"Life histories" is the term of singular importance throughout this brief assemblage of essays, in which we mean to suggest and reiterate the truth that our species has, indeed, asked every question and proffered every answer on the edge of an abyss (the Anthropocene) it alone has fostered. All of our questions and answers appear to be, and to have been, thoroughly insufficient to shed clarification on our

---

[1] "Charles Henry Turner – Scientist, Educator, Zoologist (1867–1923),"
 https://www.biography.com/people/charles-henry-turner-21302547. Accessed 27 July 2017

[2] Fjeldså J, Krabbe N (1990) Birds of the High Andes: a manual to the birds of the temperate zone of the Andes and Patagonia, South America. Apollo Books, Denmark

own anomalous characteristics, overall. The questions and answers have left us unsatisfied, indeed, as spooked as when we first began recognizing something strange, beautiful, and not a little terrifying about our predicament, all too easily written off. True, indeed: there is no reason to be re-assured that progress has occurred and that our momentary odyssey amounts to more than a fleeting glance and unaccountable accretion of unmarked graves. A monastery enshrined in every human conscience replete with the recriminations of many millennia gone awry.

But this is not to assert merely a depressing legacy or foreshadow the alliteration of the end, endlessly. Rather, it seems more than appropriate at this time in our teeming and largely tragic history to intimate something curious and timely, namely, the possibility of an individual within that collective conscience which remains a dangling modifier of human behavior and potential – a potentiality that may not be mere wishful thinking. That's the point of it all.

What do we mean, precisely? This is not a treatise that aims to foreshadow the prospects of God, although some may well choose to read of a spiritual entity or said components deeply rooted in the core of our speculations far from all the Zarathustras and Sakyamunis. Acutely more specifically, we are interested in just what a human individual might be capable of in the twenty-first century.

It would be easy enough to assume that all of the social contracts of human history, when examined in their entirety, intimate the ponderations of a species that has, however haphazardly, moved progressively forward, decade by decade, from graveyard to graveyard, ever focused upon the consecration of some special destiny unique to humans, by whatever means: sequestration, accretion, arm-wrestling, philanthropy, and disappearance. Those means and unsurprising ends include the discovery and manipulation of fire, agricultural plots, the substance and favorability of iron, mathematics, physics, the arts, electricity, the miniaturization of technology, and so forth. But these are subsets of unessential details.

The Jewish wise man Rabbi Hillel the Elder (110 BC–10 AD) "was asked to explain the meaning of Torah while he was standing on one foot. He replied, 'What is hateful to you, do not do to your neighbor. The rest is commentary.'"[3]

But for decades there has existed widespread skepticism among many scientists, but particularly biologists, paleontologists, and atmospheric chemists, that we do not have as a species what it takes to overturn that hatefulness, our out-of-control consumption, usurpation of habitat, killing of other species, felling of vast amounts of forest, and destruction of the oceans and of virtually every biome while simultaneously mounting an all-out human affront – or war against the Earth – by our sheer proliferation of consumers. A hideaway in Scotland, such as that intimated above, is a temporary respite from the written litanies of human history, as are Woody Allen's finest one-liners.

But the collective assault on the planet's carrying capacity threatens a "replay of the Paleocene-Eocene Thermal Maximum (PETM) of 55 million years ago, risking a very likely catastrophic end to global civilization."[4] The PETM represented a

---

[3] Morrison R (2017) Sustainability sutra: an ecological investigation. SelectBooks, New York, p 1–2

[4] ibid., Morrison, p. 109

200,000-year-period in which excessive amounts of carbon dioxide were infused into the atmosphere and troposphere, creating a cybernetic hell for a large number of terrestrial organisms, a disaster mirrored by our present surge beyond the 400 parts per million of CO_2 injected by our ever-escalating industrial paradigms, into the atmosphere.

But the primary fuel of this ultimately biological crisis is a deep demographic, a mindless fertility rate, and a built-in population explosion most recently reflected in a lawsuit filed in federal court by the Immigration Reform Law Institute against the US Department of Homeland Security "for failure to properly comply with the National Environmental Policy Act (NEPA)."[5] The essence of this emblematic lawsuit hinges upon the fact that NEPA (instituted on January 1, 1970) requires of each and every US federal agency environmental accountability, and claimants most assuredly include individual human beings, whose environment may be affected adversely by a federal action. According to Leon Kolankiewicz, "America's total population is projected to increase to 441 million by 2065 – an increase of more than 115 million from our 2016 population. [And] Demographers estimate that immigration will account for 88 percent of this growth."[6] This, in turn, will likely impact indigenous biodiversity and habitat by "1.2 to 2.2 times greater than they are [impacted] at present."[7]

Lost in such a lawsuit is the global picture. Human-imposed borders meant to define specific sovereign states are not only obsolete but delusional. Carrying capacity is not an exclusive right. It is the greater law of natural duty that transcends borders and the naming of groups of people or other organisms in an ultimately doomed effort to ignore them, obfuscate them, and reject them. Migratory species wisely, though often at their peril, have no patience for the nonsense of human-imposed boundaries.

In all the realms of ecology and individual accountability, every individual most assuredly counts, certainly qualitatively. Meanwhile, quantitative analyses underscore the historic differences and long-standing gulf between notions of "individual" and of "community." These are embedded biological concepts that enter the realms of fuzzy logic, without any strict definition that owns up to the many realities swirling around the words and the feelings they evoke. Typically, when we try to fathom the individual, we impulsively invoke *individual freedoms*. But freedom from what? Freedom to do what? Move to a relatively isolated Scottish wilderness retreat? Whose freedom? That of a community of communities or just particular communities inhabited by specific individuals? Connected or unconnected communities? Communities in turmoil or concord? Dependent or independent of each other? Communities that survive according to the classic (deficit-logic) "Netherlands fallacy," "broken windows fallacy," or other natural capital depreciation syndromes, from Javanese soils to disappearing wetlands in California – in other words,

---

[5] ibid., Morrison, pp. 1–2

[6] See "An Environmental Impact Statement on U. S. Immigration Levels," by Leon Kolankiewicz, CAPS ISSUES, January 2017, p. 1

[7] ibid., p. 3

communities that have breached their carrying capacity? Or communities wherein the Aristotelian polis ensures a sustainable collective of biological habitats that serve uncompromised, and unviolated, the fullest needs of the entire array of species whose homes are encompassed by individuals who, in turn, must own up to the obligations that come with environmental citizenry?

With the human species, we necessarily recognize at once an enormous problem. By *human species* within a community, there is but scant evidence (from ethnographic sources, many arcane or of an earlier era – e.g., not a single papyrus that survives the world of the Canaanites) that conveys anything like intergenerational equity, gender parity, and individual case-by-case, household-by-household equality. In other words, the minute we invoke the species category, we are erecting barriers to individual rights that equate with what has been described by many philosophers and legal analysts as *distributive justice* or injustice. This represents a moral minefield that is like some geological great divide.

We see the profiles of that abyss articulated and/or intimated by such philosophers as John Rawls (1921–2002) and John Stuart Mill (1806–1873). Rawls' book *A Theory of Justice* (1971) never mentions "community," though in his defense of a social contract, he constantly references "society" – a fact that segues to a situation whereby "people somehow aren't aware of their circumstances and therefore don't know whether or not they will benefit or suffer from a decision."[8] Conversely, Mill fundamentally considered "the relationship between freedom and community."[9]

Both sets of deliberation – the isolated individual and some working collaborative co-symbiosis or at least commensalism – invite proposals for evolutionary success or failure, justice or unfairness. Neither game plan has ever assured anyone of a biological status quo. Nor has "anyone" ever translated into "everyone," let alone everyone of every other species. At least in the polymath of John Stuart Mill, we see a man who engaged in "conversation rather than pronouncement."[10] But only among those rarified pantheist ethical traditions (e.g., the Jains and ancient Taoists) has attention been significantly paid to all species and all individuals of those species.

That focus has been largely ideological. Gandhi, Mahavira, Christ, Buddha, Lao Tzu, the Essenes, the Tasaday of Mindanao, the Inner Badui of western Java, the Bishnoi of Rajasthan and Todas of Tamil Nadu, the Sufi vegetarian Etyemecz neighborhood of Istanbul, the Karen of Myanmar, the Hadza of Tanzania, the Chalingpas of Bhutan, and organizations like People For the Ethical Treatment of Animals have all espoused this pantheism, dating back to poignant philosophers like Thales and, much later, Leonardo Da Vinci and Percy Shelley. Many of these ambassadors of non-violence have indeed spawned whole movements, religions, revolutions, and grand paradigm shifts that involve the capacity of individuals to make a difference in larger spheres of influence – the ecological David versus Goliath.

---

[8] op.cit., Morrison, p. 81

[9] ibid., p. 83

[10] Colin Heydt, "Internet Encyclopedia of Philosophy," 3. Conclusion, http://www.iep.utm.edu/milljs/Accessed 28 July 2017

The very rudiments of every ecosystem are forever in flux. Species like ours come and go with the mental sense of a great acceleration or unforeseeable collapse of all that we hold essential and dear. What rings of all too familiar clarity are the peril of tenuous times and the chemistry of sleepless nights. That's who we are as a collective. What are we as individuals? *Are* we individuals? The question means to split wide open the obvious bias of countless millennia.

No one has ever outsmarted destiny or outrun their own shadow. We have lauded every laurel branch upon our narratives (Leonardo's "Ginevra de' Benci" comes to mind) and engendered Baroque frames to embroider the landscapes we inhabit, cropping and editing our story so as to possibly elude the clear and present evolutionary dead-end that is every day more likely. Its likelihood is founded on our self-interest.

Do pessimists dream? Or have they already gathered in the ruinous spillway of their prognostications, awaiting a new dawn that will never come? Separating oneself from the masses (the subject of Thomas Hardy's 4th novel), the oft-cited injunction that has galvanized generations of artists and freethinkers, does the individual have the strength and endurance, whatever it takes to singularly impress upon a crowd, a number of souls far vaster than herself/himself, some new vision of a more viable nature than human history has thus far illustrated?

This query can be lodged in any number of ways, but comes down to a simple, if circumspect, conundrum of our annals: How can one influence many? Is human evolution alive and well and, if so, is it favoring individuals or the species? Individual "fitness" or inspiration? Individual survival or surfeit? Compassion or greed? A superego or some regenerative humility?

Can a future individual from the human species significantly engender a force for nature that is effectively team-oriented, non-violent, and nonexclusive? Or are the sum total of evolutionary rubrics working at a pace and style that undermine a revolutionary individual's contributions and influence over her/his species? Is an ecologically hybrid future, as it has come to be understood by the biological sciences, in our best interest? Hybridization connotes a combining of circumstances in which the power and reach of known human industry are vested to varying degrees to cushion the individual from calamity. The agents of its propagation are enriched through law, custom, contract, and force. All devolve from the natural sciences and human nature. These are the community charters akin to biological speciesism, whereas the individual's actual independence, integrity, and influence must combat a system of supremacist feudalism in all of its continuing iterations – a condition akin to the brutal hierarchies we perceive in most social insect societies. In fact, our belief in self, and self-rule, has little wiggle room in the actual annals of biology. We are, it would seem, as fixated as any worker ant, despite equally stubborn instances of helping hands, Arcadian retreat, and artistic expression.

Ultimately, we must ask: What are the sociological and medical dynamics that favor individualism over populist genomes? We speak frequently of individuals getting off the grid (the electrical dependency upon conventional means of acquiring kilowatts); but what about the prospects for individuals separating themselves from mindsets and likely outcomes of a geopolitical, legal, and economic grid that has, to date, enforced nothing less than acquiescence and outright serfdom to the multinational and federal powers that be? Much like the serfdom of women in Victorian England (subject to virtually no freedom, whether within or outside marriage). If a freethinking rogue has somehow gotten elected or been chosen within the Beltway, it is clear that Beltway was already inside that rogue.

Is there a hypothetical individual worth conceptualizing whose place in the hierarchy of power defies any stochastic target, mathematical certainty, or moral ambivalence? A personage whose mental and ethical infrastructure is beholden to no hegemony other than the globalization of empathy, shared by many, as it has largely been eschewed by vaster human numbers, and whose dedication throughout any given life cycle recommends the possibility of a human individual who will see us past the eschewal, the Anthropocene, acting presidentially in the most modest of gestures, seeing to an anthrozoological orientation that reveres all those other Individuals, from among the incalculably rich Tree of Life? In this same vein of hope and redress, George Steiner intones, at the beginning of his rich text on meaning and transcendence, the declarative, "One of the radical spirits in current thought has defined the task of this somber age as 'learning anew to be human.'"[11]

In this brief and mostly generalized treatise (we feel not the inclination to render an encyclopedia in place of a few, hopefully salient or at least relevant observations), we have chosen some case studies – individuals, works of philosophy, natural history, anthropology, paleontology, the ecological sciences, comparative literature, ethics, and spirituality – to better grasp what such an individual might look like; how she/he might behave; and what motives, aspirations, and methodologies such an individual would be likely to embrace in order to make a profound difference for life on Earth, one favoring perpetuation and biodiversity, non-violence and love, over any contrary forces. As the lead paleontologist for the Jebel Irhoud research team, Jean-Jacques Hublin declared, "The story of our evolution over the past 300,000 years is mostly the evolution of our brains."[12]

Our choice of essays, and their cumulative approach, is as personal as it is (hopefully) instructive.

Los Angeles, CA, USA   Michael Charles Tobias
                                                                                Jane Gray Morrison

---

[11] Steiner G (1989) Real presences. The University of Chicago Press, Chicago, p. 4

[12] "Humans evolved 100,000 years earlier than thought and East Africa is not 'cradle of mankind', say experts," by Sarah Knapton, Science Editor, June 7, 2017, http://www.telegraph.co.uk/science/2017/06/07/humans-evolved-100000-years-earlier-thought-east-africa-not/. Accessed 8 June 2017

Contents

1	**What Does Humanity Mean?**	1
	Species and Individuals: Two Narratives at the End of Days	1
	Wittgenstein's "Tractatus"	8
	A Raison d'être	13
2	**The Lost Tribes of Tamaulipas**	17
	A Primordial Consciousness in Northern Mexico	17
3	**Edward Curtis' Vision of Transcendence**	27
	North American Ecological History Leading to Edward Curtis	27
	Archaeological Evidence of the Cultural Vortex that So Inspired Edward Curtis	28
	Mixed Affairs	30
	When Strangers Meet	32
	North American Indian Demographics	37
	Enter Edward Curtis	39
	The Revelation in Seattle	43
	Curtis' Great Irony	46
	The Rembrandt of Photography: Edward S. Curtis and America's Environmental Social Justice Movement	47
4	**A Genetic Cul de Sac**	49
	How the Individual Reshapes the Species	49
	Genetic Fitness	56
	The Coyote Conundrum	58
	Latitude or Cul de Sac?	63
5	**The Individual Versus the Collective**	67
	The Vicissitudes of the Self	67
	Beethoven, Darwin, and Parmenides	69
	Psychiatric Genetics	75
	Reclaiming the Individual amid Speciation	80
	The Primacy of Gauguin's Coda	84

6	**The Separation of Mind from Species**	89
	Ethical Cognition in an Age of Despair	89
	Beyond Self	94
	Biocomputing the "Personage"	96
	Amidst or Among?	97
	New Mind, New Infrazoology	100
	The Individual Amid Multitudes	106
7	**Ecological Existentialism**	109
	The Poetry, Science, and Despair of the Anthropocene	109
8	**Choices**	121
	Jerome's Choice	121
	Ontological Madness	122
	Vexatious Variables	125
	"Readiness Potential"	127
	A Psychiatric Black Hole	131
	Judgment	136
	Not Who but What?	138

About the Authors

Michael Charles Tobias earned his Ph.D. in the History of Consciousness at the University of California–Santa Cruz, specializing in global ecological ethics and interdisciplinary humanities. He has conducted field research in over 90 countries, producing a wide-ranging body of work that embraces the history of science, aesthetics, anthrozoology, comparative literature, philosophy, and natural history in the context of a multitude of current and potential-future scientific, geo-political, economic, and social scientific scenarios. Tobias has been on the faculties of such colleges and universities as Dartmouth College, the University of California–Santa Barbara, the University of New Mexico–Albuquerque, and Georgia College & State University. For 18 years, Tobias has been president of the Dancing Star Foundation (www.dancingstarfoundation.org; www.dancingstarnews.com). *The Theoretical Individual* is Tobias' 4th book with Springer.

Jane Gray Morrison executive vice president of the Dancing Star Foundation for 18 years, has written and co-edited dozens of books, including *Sanctuary: Global Oases of Innocence* (www.sanctuary-thebook.org), *Donkey: The Mystique of Equus Asinus* (Council Oak Books), *Why Life Matters* (Springer), and *Anthrozoology* (Springer). In addition, Morrison has written, produced, and/or directed numerous major film documentaries and docudramas that have been broadcast throughout the world. Among them are the ten-hour series "Voice of the Planet" (TBS), "A Parliament of Souls" (PBS), "No Vacancy" (PBS), "A Day in the Life of Ireland" (PBS), "Mad Cowboy" (PBS), and "Hotspots" (www.hotspots-thefilm.com, PBS). A goodwill ambassador to Ecuador's Yasuní National Park and early advocate for an Antarctic World Park ("Antarctica: The Last Continent" [PBS]), Ms. Morrison has worked to save endangered species and habitat on every continent. Those efforts have included the creation and management of one of the most successful mainland island scientific reserves in the Southern Hemisphere, on Stewart Island, New Zealand.

Chapter 1
What Does Humanity Mean?

Species and Individuals: Two Narratives at the End of Days

One may assume that we know well what an individual is. That we cherish her/him, though not so much, it; that we confer rights, dignity, esteem, a special place for individuals. That we can readily ascertain the difference between an individual and a species (a matter of linguistics, and of course, a variety of numeric and genetic calculations, if not mere counting). That the entire history of science and of the arts has ennobled individuals, giving way to a pluralism that fancies any number of pantheistic scenarios. But these are the tendons of a pathological falsehood, a self-perpetuating premise that is as devious as a card trick and as ruthless as an assembly line of chickens destined to be slaughtered.

Billions of chickens murdered for our pleasure every year are not viewed as individuals, but, rather, a lump sum of profit for those engaged in the business of slaughter and all its ancillary distributions of colossal injustice and horror.

So let us begin by trying not to kid ourselves. Nothing is as it seems. The most blatant, glaring, obvious reality – that of an individual – has been blurred by all that with which science and human beings have surrounded it.

In the ebullient, controversial, technically proficient physician and always surprising philosopher Julien Offray de la Mettrie's (1709–1751) book, *L'homme Machine* (*Man a Machine*,1747) the raconteur of humanity's evolution writes:

> What was man before the invention of words and the knowledge of language? An animal of his own species with much less instinct than the others… he lisped out his sensations and his needs, as a god that is hungry or tired of sleeping, asks for something to eat, or for a walk. And nearly concluded, Let us not say that every machine or every animal perishes altogether or assumes another form after death, for we know absolutely nothing about the subject… [and speaking of butterflies] The soul of [these] insects (for each animal has its own) is too limited to comprehend the metamorphoses of nature. Never one of the most skillful among them could have imagined that it was destined to become a butterfly. It is the

© Michael Charles Tobias and Jane Gray Morrison 2018
M.C. Tobias, J.G. Morrison, *The Theoretical Individual*,
https://doi.org/10.1007/978-3-319-71443-1_1

same way with us. What more do we know of our destiny than of our origin? Let us then submit to an invincible ignorance on which our happiness depends.[1]

That we have not submitted to anything like "an invincible ignorance" can be easily gleaned from a compelling overview of humanity's orientation to deep ethology, the science of our species in relation to others, as elucidated by Dr. Michael Giannelli in 1985:

> We must remind ourselves that scientific progress is not invariably human progress. The continued expansion of human knowledge at the cost of human character is a pathetic trade-off which, if continued, will eventually destroy the civilization we glorify. At present, there appears to be decreasing prospects that humanity will ever make peace with itself. Many say, this being the case, how can you expect humanity ever to make peace with the rest of the animal kingdom? …other animals do not represent the threat to us which we do to each other. As a species, with good reason, we distrust and fear each other far more than we distrust and fear other animals. The only animal which threatens to push us away from the dinner table is man himself. Perhaps, just perhaps, there is hope for facilitating the peace process by first de-escalating our aggression against the other, less warlike, species which inhabit this fragile earth.[2]

These two narratives, separated by nearly 240 years, encompass most of the issues persistently challenging the modern history of animal protection insights and activist movements; of the emergence of genetics; Darwin's and Wallace's theories of evolution; an international animal liberation movement and concomitant neurophysiological research into diverse species' interrelations and interdependencies, languages, behavioral empathy, or lack thereof; the whole realm of neuroplasticity and its memories, impulses, fear, hope, pleasure, and pain centers in the brain.

In other words, as underlying texts for micromanaging our species' global predicament, between La Mettrie and Giannelli, we are confronted with a striking latitude of liabilities implicating consciousness; the beginning and end of science as it is instinctively applied to our relations with, and perceptions of, what we have long termed the Others.

On August 5, 2010, a Google algorithm suggested there were 130 million separate titles of books that had been published, as of that date, throughout the world. We cannot fathom the remaining number of unpublished manuscripts, journals, and diaries, but all this speaks – in addition to every other form of human expression – to our irrepressible needs. Herewith we add yet another tinder to a gigantic – what? The span of time, of course, Gutenberg to Google, is merely some 550 years or so. Additional paleontological forensics grants our species another 200,000 to

---

[1] Shalizi CR Cosma's Homepage. http://bactra.org/LaMettri/Machine/. Created 31 Mar 1995. Accessed 7 Oct 2016; See also http://www.earlymoderntexts.com/authors/lamettrie/. Accessed 7 Oct 2016; See Philosophical And Historical Notes By Gertrude Carmon Bussey (1912) Man a machine, including Frederick The Great's "Eulogy" on La Mettrie and extracts from La Mettrie's "the natural history of the soul". Open Court, La Salle

[2] See "Three Blind Mice, Michael A. Giannelli (1985) How they run: a critique of behavioral research with animals". In: Fox MW, Mickley LD (ed) Advances in animal welfare science 1985/86, The Humane Society of the United States, Washington, DC, pp. 109–164. http://animalstudiesrepository.org/cgi/viewcontent.cgi?article=1007&context=acwp_arte. Accessed 4 Sept 2016

330,000 years, approximately. What can we definitively allude to and about ourselves that has not been declared, adjusted, calibrated, or divined before? Probably nothing. Yet, in nothing, there is always something we hope. This is not meant as perplexing nonsense, though that would certainly explain the impulsive repetitions that are seen throughout the natural and unnatural world. All repetition follows a variety of laws that are both physical and biochemical, from the essence of templates and subsequent replication in cellular biology to the nature of frequencies and waves in all acoustics. Gravity is most assuredly repetitive, and the adherence of an apple – the famed Flower of Kent – to that compulsive law of the universe, it is imagined, is but one lovely example that every Tenzing Norgay on his Everest can relate to.

If we have heard it before, as a species, whether in the case of John Ruskin in his four essays published together as *Unto This Last* (1860) citing Matthew 20 from the King James Version of the Bible, to a proliferation of Golden Rules and yet other bibles, we still seem to need, as the great naturalists Henry Beston and Alexander Skutch accordingly have recognized, to be reminded again and again to look closely at something, until we finally see it. It is common sense, but in natural history, *sense* promotes unexpected comparables and corollaries that defy generalization, and by frequencies, color spectra or some other element of physics, physiology, genetics, or chemistry are simply out of our species-specific grasp. This is basic. But the impact of ecological illiteracy, or exhaustion, or simply the indifference that arises at any number of life's junctures, most probably weariness and the sheer weight of trying to hold on, is a universally valid argument leading us again and again through the same dark tunnel, toward a uniformly metaphorical light that offers us no bargaining. We will not grasp it. Ever.

What, then, are the prospects for a humanity with 130 million book titles to its credit and more woes in the early twenty-first century than ever before in its brief history? We grew up in America hearing that progress was slow, erratic but inevitable; that every facet of social injustice was on the mend. As children we come by turns to know the fickle sources and explosions of injustice, even if some of our realities may be inordinately sheltered compared to the phlegmatic veneers occupied by others of our kind. For many – a scathing truism borne out by every international agency statistic – nurturance is not a given or certainly not for very long. And protection, even less so. When we are tired, frightened, or in pain, it is tempting to grab hold of evolutionary theories with vigor, for such notions seem to speak personally to us and explain our misfortunes. There is every reason to seek solace on Earth, and there lies a vast edifice before us speaking to countless bioreveries that have preceded our time, from ecstatic cave paintings to Impressionist Picnics. But there is even more fodder when we look out upon harsh combat in so many forms, whether between nations, genders, or carnivores. A story line has built up this biology of war; a psychoanalysis of discontents driving a statistically significant "ecology of sleepless nights," this latter phrase from one of those 130 million titles, *World War III: Population and the Biosphere at the End of the Millennium*.[3] Bach versus the Holocaust; early human cannibalism (e.g., Maori cultures devastating

---

[3] By Tobias MC (1994) Bear & Co., Santa Fe, NM. Revised 2nd edition, Continuum Books/Herder & Herder, New York, 1998, with a Foreword by Jane Goodall

earlier Moriori habitation on the Chatham Islands [New Zealand waters]); the Tasaday of the late 1960s, their champion most notably, John Nance, as opposed to subsequent obfuscation by politically vested interests; a long history of vegetarianism, veganism, and animism stacked up against the seemingly implacable PCP (palate convenience paradigm). We are a species of dialectics and of two underlying predicates, opposites, and archetypes: Good and Evil. These are generalities, of course, that apply for most purposes that presume some Theory of Knowledge.

Whereas ecology, with its myriad disciplines, is actually quite personal, with countless empirical and theoretical pursuits converging zealously in scientific and cultural realms to enrich its outreach. Personal outreach. We may count the number of lemmings who perish cyclically every 3 or 4 years in many northernmost regions of the planet, the boom-and-bust demographics that challenge the survival rates of every species. The most specific research on this ecological fundament of the food chain comes from the predator-prey relations studied with respect to cougars atop the North Rim of the Grand Canyon in the 1930s; but that same metaphor extends most cruelly to those Boreal lemmings – only *H. sapiens* seemingly having thus far wriggled free of such cycles. But there is little *personal* comfort in such numbers. Personal comfort, however, is what such demographics ultimately relate to. We see the face of individual lemmings, not just a mob of lemmings. Again, there is inherent to this span of time, this index of actuarial casualties, a clear line that moves steadily in the name of science, just as the varied "arts" have overwhelmingly sought beauty and insight.

Theoreticians may take some pride in elegant theories and even shed tears over General Principles, as much as any audience at the end of Mozart's "Marriage of Figaro" is also likely to do. But our ability to shed tears is a very mixed bag in terms of lessons and consequences for our species at large and for the planet. This much we may easily adduce from history. We do not easily acknowledge a universal theory of tear ducts in other species that has demonstratively changed our behavior toward them. "Elephants weep"; dogs – according to zoologist Henri Coupin and John Lea. *The Wonders of Animal Ingenuity*, Lippincott, Philadelphia, 1910) – not only whimper and howl, but also cry; fish remember everything, sea otters mourn their lost loves, salmon, penguins, and albatross return to the spot where they were born. Dogs, burros, cats, manta rays, orangutans, and so many others, each and every one of them, convey their love, altruism, and wisdom.

What is far more difficult than any theory we may arrive at, and underlies the heart of our present concerns, is to what extent humanity is capable of being good, pure, and simple. Of truly loving. Such sentiment may strike at first glance somewhat childish, in terms of repeating old adages and beseeching worn-down quests. Is not humanity, the subject/noun, implicit with meaning steeped in goodness? We would all like to think so, especially at those most horrible of times when "crimes against humanity," for example, are invoked. Nor do we disagree with the fact many, many individuals are good, notwithstanding the pressures upon them to simply get through each day and night; the many roiled and wheeling rages toward efficiency, survival, that easily militate against that greater charity in our hearts. Following orders is one of the great examples used to bridge this underlying dialectic (particularly evinced in the Nuremberg trials).

Does that charity, the notion that humans are basically good, actually speak on behalf of our species? And if so, how, then, has *H. sapiens* managed so ruthlessly and collectively to unleash the Anthropocene, the sixth wave of known extinctions in the history of life, but the one exhaustively fulminated as the result of our sole species' behavior? This is the question that plagues us, quite literally tormenting and realigning all alleged torments within ecological, ethical, legal, social, and a bewildering array of other disciplines and contemplations.

Most ecologists and people of spiritual longing have rather tacit rejoinders to, but few if any overall solutions for, this grim vexation, calling into earnest doubt the reliability or stable infrastructure of our behavior. And as we shall explore in this thin meditation on the topic, most of our crises actually offer up no instructive nor applicable mitigations. Philosophy and Utopian imaginations have had to suffice for a demonstrable want of any so-called solutions. You may well come to that conclusion, as well; or you already have. In coming chapters we'll offer our own opinions on this topic.

Who we are as *individuals* is somehow separated, more and more, from *what* we are as a *species*, part of the critical point the indomitable Lord Chancellor Francis Bacon (1561–1626) was making near the conclusion of his monumental thought experiments and observations upon natural history. If a convergence were possible, one would assume we would have had enough time and practice by now to get it right. We have not. One might also argue that it would not serve goodness for the minutely differentiated, nearly seven-and-a-half-billion of us (and quickly increasing in numbers, still) to become a uniform force. That might easily backfire in one particularly galling and immediate manner, namely, our total self-destruction, the result of an unquestioning and monolithic choice made on behalf of our entire species; a single mood swing by someone with a sufficient bulwark of power to ignite a nuclear war, to take the most obvious example.

Whereas our varieties of experience and genes, our proliferated handprints, make of every different cave wall a new and possibly wonderful illumination that works against the outbreak of Hitlers and Pol Pots and others.

Yet, despite declarations of hope and courage, this schism persists with aging vehemence. The *royal we* are destroying Earth, challenging our alleged humanity to somehow make a difference in time to save the majority of individuals, both among humans, but even more pressing, among the Others who are so fast vanishing. Seen as a moral crisis of the moment, but lacking the "boldness" and "genius" Goethe referenced in his Faust, we may well find that our species is capable of analysis in the absence of an actionable momentum. Lingering in the philosophical purgatories of our wish fulfillments, the paroxysms accompanying the evolution of individual departures from the norm – across a time frame of tens of millions of years – depriving us of the catalyst that might yet richly infect a stubbornly inward-dwelling species, we stagger and foment atop the lost cliffs of Kurosawa's "Ran," his haggard and betrayed daimyo/King Lear forcing his face into the gusty Eumenides, incapable of imagining family reconciliation.

Biophilia-longing intellects have no greater purchase upon the destiny of desolation than all those populating the battlegrounds that encincture their paradise (sic). Chance has stripped them of any means to achieve sobriety and calm, as the composer of "Ran," Toru Takemitsu, made so immeasurably clear.

How do individuals accomplish the mighty hat trick that is repose? Is it possible, or, as so many have perennially alleged, are we doomed as a species? A doom that is, by present, overwhelming evidence, far in excess of ourselves.

If by preference, order, logic, and tiers of premeditation we undertake to pick apart this atrocious conundrum in an arguably tenuous attempt to right wrongs on paper, we are, admittedly, adding to the exasperating fault lines of past tirades, religious divining rods and repetitive discourse. A tired rhetoric, belabored, and almost instantly obsolete, in an age succored on instantaneous image, word, gratification, sound bites, and 3-D printing. That rage for the immediate signals is a clear warning that we should be concerned about the future, where no one prior to our time appears to have given it a thought (with the possible exception of those who designed Chartres Cathedral, which required hundreds of years to complete and thus, one adduces, instilled in its multiple labor forces a multi-generational commitment to some hazy future). Yet, the ground beneath our feet has always felt safer, presumably, than the horizon line. That the miseries we know are safer than those we have yet to experience. These harken back to ancient Codas but take on particular poignancy at a time when the economic fallout and psychic wounds entrapping humanity risk the absolute fatal distraction: a solipsism that has, of necessity, turned its back on the last chances to engage in that goodness which might yet portend of a renaissance of virtue and a literal emancipation of all those Others who are, by our callous withdrawal into ourselves, lost forever. We refer to those trillions of other animals and plants that will disappear, and not before their long and unimaginable individual suffering.

The distractions masking this chatter above the ashes are comprehensive and all speak to us, not to them. What was narcissism is now the new normal. We have heard often eloquent rants against greed for at least three millennia; been swayed by musical Messiahs, and others, whose beauty or persuasiveness has occasioned global day offs, like Christmas. But we have not been swayed the day after.

So these, then, are the stakes that humanity has thus far failed to take seriously enough: a global extinction, the very sounds and thoughts of which have actually become one more distraction, the ultimate irony for all those societies which have become sick of duty, disdainful of all authority, mistrusting of human nature, fed up with the collective status quo, and individually resigned to it, all at the same time.

The grandiloquent perorations of inaugural addresses, judgments from elders and supreme courts, all of the ancient wisdom, have ignited in this generation a fallacy expressed in anger and a sullen medical condition that Pieter Bruegel most likely endured across a landscape of perpetual warfare and exile; the behavioral quotients of that universal distraction which all evolutionary thought has recognized as one unshakable impediment to a kind of transcendence assumed to be biologically impossible, in any case, ethically and intellectually incomprehensible, namely, the *merging of separate species.*

We don't know what that means; we can't imagine what it means; we don't want to think about that, save for the vaguest of analogies within the physiology of tree reproduction when suckers are involved, from elms to cottonwood, poplar and western redbud.[4] It enforces a noxious scheme of things that so unhinges the order of the Cosmos as to force us into those woods teeming with the witches hovering around Macbeth, where "Fair is foul, and foul is fair."

Aesop and Attar convened parliaments and conferences of birds, but each bird was of a separate species. Noah's Ark and all of its preliminary and subsequent commentary has singled out each and every type of being and decreed that, for the proliferation of their specialties, there be two of them, in the mid-eighteenth century lending a credence to binomial nomenclature with the firm fundament of sexual reproduction. But there were other pillars of constancy, as well; feathers filtered through plumes that lent to feathers both their exquisite fineness and a diversity of color versus hair through the skin that has never come anywhere close to the proliferation of color types among bird plumage, as Bacon had endeavored to conceptualize.

Our myths are more powerful than we realize. We see the world through them, and mythology does not sanction, for example, the building of a fast-food restaurant atop Mount Kailash (Sumeru), a mountain in southwestern Tibet sacred to more people worldwide than any other point of high altitude. And while Kailash remains as yet inviolate there has already been the first Chinese road survey nearby and this level of engineered denigration is precisely one sort of path upon which humanity finds itself.

The myth of humanity is etymologically, psychologically, and emotionally ingrained in who we are. Yet, every minute somewhere it is violated by humans. We have asked enough times. Why do we persist in betraying ourselves? In asking that, we seem to assume to be asserting certain pure origins. The Chateaubriand-driven myth of the noble savage, for example. For anyone who has lived months or years unbuttressed in the jungle, the desert, in a cave, on an isolated island, or at high altitude, to name but some of the original architectonics of a vastly reversed global human population, there are a host of discussable indelicacies that arouse less the curiosity than the forbidding problems attendant upon isolation from comforts and some level of security. Now it can be added that the life of a monk these days is easier than that of monks during, for example, the period of Emperor Justinian, where slaughter of religious mendicants was frequent. There are few saints whose lives did not end horrifically.

It is also clear that until their fame spilled upon the world, first in the 1840s, the pre-Davidian Todas of India's southern state of Tamil Nadu and, then in the late 1960s, the Tasaday of the island of Mindanao in the southernmost Philippines, speaking a unique dialect of Manubo- part of the Malayo-Polynesian language group,[5] both of these tribes, numbering fewer than 1300 (the Todas) and 100 (the

---

[4] Flanigan A, SFGate Plants That Reproduce From Suckers, http://homeguides.sfgate.com/plants-reproduce-suckers-25699.html. Accessed 4 Aug 2017

[5] See Nance J (1981) Discovery of the Tasaday, a photo novel: the stone age meets the space age in the Philippine rain forest. Vera-Reyes, Inc. p VIII

Tasaday), may well have inhabited two distinct paradises, as our mythologies tout such Edens. But few are the Gauguins, Wilfred Thesigers, John Claude Whites, Francis Kingdon-Wards, or Thor Heyerdahls who have, even for a few years, managed to *go back* to that paradisiacal nature.

Humanity comes as loaded with baggage as that fleeting, grasping conjuration that the *grass is always greener*....With that as given, at least, we can take some stock of our compass reading and weigh in. Early twentieth-century philosopher Ludwig Wittgenstein certainly did so.

Wittgenstein's "Tractatus"

In his 1921 publication of *Logisch-Philosophische Abhandlung* (*Tractatus Logico-Philosophicus*, the British/American 1922 edition, translated by Frank P. Ramsey and C. K. Ogden), Ludwig Wittgenstein formulated a methodical cipher, delivered in a series of *declaratives*, whereby human language (at least the German and English in which he delivered the first two editions) might be seen as a fitting simile for unlocking the meaning of all human languages (in his time that meant well over 7000 of them) and of *meaning* in general.

This encompassing approach to our lives, to all of the questions at the core of philosophy, metaphysics, pataphysics, and the ecological sciences, has unraveled amid a fanfare of insulting and heartbreaking chaos that so engulfed the twentieth century, lending a particular poignancy to Wittgenstein's choice of writing style (terse and enigmatic) and underlying goal – the attempt to somehow convey that which cannot be conveyed and thereby shed a spotlight on the ineffable in human nature, language, and the meaning of meaning.

Quite inadvertently, he also opened up a metalanguage for investigating the communication skills and illuminated lures among invertebrates, particularly those insects in high- or low-altitude migrations, research of which has only accelerated since Wittgenstein's time. In just under 4 h at the Santa Fe Opera during the 2017 Summer season, during a brilliant performance of Handel's "Alcina," the positively phototactic excitation rate and organismic density within an approximately 100 cubic feet area of illumination must have exceeded – by the authors' count – several hundreds of thousands of inspired insects. These included the red-eyed *Rugusana* leafhoppers, coneworm, and twirler moths,[6] among dozens of other species. No one has yet come close to counting the individuals taking part in such migrations – unlike, for example, wildebeest. Finding ways to establish a baseline for insect attraction to bright lights, let alone opera, noting that at least half of all known insects can hear, poses its own subjective, metaphysical challenges, largely of the imagination. Once while returning by small outboard motor craft to a research camp in southern Borneo, our large lantern, stationed at the helm, revealed in the wide

---

[6] See "Bug Eric," http://bugeric.blogspot.com/2015/12/new-mexico-night-bugs.html. Accessed 4 Aug 2017

glare of an encinctured remote river what appeared as millions upon millions of insects and some bats. The continuous throughflow of insects across the planet translates into a flying biomass that might readily be described as the second most populated ensemble of language groups on Earth, after uni- and multicellular organisms. All these *individuals* must reawaken an entirely new reading of the *Tractatus*, and the duties and moral obligations we humans are necessarily confronted with. But most charismatically, the cellular and entomological worlds excite a biosemiotic fundament that changes everything we can possibly interpret, dream, or cognize.

One assumes that Wittgenstein intended by his deceptively terse expanse of the *Tractatus* to establish the reality of millions of languages, not merely the 6500 or so remaining known human ones, not including, according to neuro-linguist Marc Ettlinger at UC Berkeley, some other "4000" human languages that have probably gone extinct.

Yet, upon the written page, there is little to indicate whether Wittgenstein was the least concerned with other species' languages, as he moved inexorably and logically toward a universal silence so at odds with what we take to be a realm of biological proliferation crucial to actually beginning to understand the world in which we live. In Wittgenstein's quest to establish the pillar of all human philosophical determinism, a phrase that cannot be separated from what is, after all, many tens of thousands of years of cramped and cringing human introspections, he invented a plurality of questions without a single governing framework of answers. Stated differently, Wittgenstein established a philosophical vacuum, the actual *absence* of any meaningful links between the world and ourselves as a means of splendid and perplexing isolationism. He proffered a universe where there is no evidence for the necessity of a single question or the acknowledgment of aesthetics, logic, coherence, or, for that matter, incoherence. In the *Tractatus*, no prima facie trappings of necessity adhere to a single subatomic particle. Nor do any laws of physics or chemistry or biology necessarily apply to human thought. Wittgenstein's masterwork illuminates our loneliness, our liberation, our hopelessness, and even our rationale for being indifferent to fate.

If some of us (not just historians) care about all those nameless millennia and anonymous considerations rushing to judgment, then we must wonder what significance, if any, that contagion of restless, meandering deliberation entails; the act of making history, of finding ruins, divining principles, engendering rhyme schemes or librettos like that written for an "Alcina" devoted to a magical realism dating to Homer, Ovid, Shakespeare, and Ariosto in which individuals and species were seamlessly merged and translated; the naming of names, keeping timelines to better contextualize the emergence of Egypt's pyramids, or the invention of penicillin. Implications are merely a second out of time, a once removed layer of uncertainty, confirmed by the truth of all the largely forgotten lives of our forebears or descendants to be. We inhabit a Cosmos which we perceive from a narrow and adverse blind, and nothing we think, feel, or do has ever shown the slightest tendency to change our overall situation. We are, however, enthusiastic or inspired, working from a dark bunker, no matter how many constellations we may have named to aid us in our ephemeral navigations.

Wittgenstein's method throughout the *Tractatus* deploys a rhythm of demonstrative assertions steeped in a *music* of conceptual finality, of statements arrayed in some order that is meant to be inculcated as fact and presumed to have successfully defined and shaped the entire sphere or analytic geometry of human boundaries. We doubt if that is the case. But we don't know.

The many extreme peripheries of Wittgenstein's coordinates are likened to something implacable in the existence of the world. Like x or y, of their quadratic, logarithmic, or exponential quantifications. So, for example, categories long debated by philosophy, such as "form," "content," "existent," "object," "variable," "existence" and "nonexistence," words pinioned on the map of the world by Wittgenstein, are each differentiated by a peculiar function, both mathematical and psychological, namely, the omniscient narrator (the Wittgenstein in each of us), who is both within and outside his own theoretical human nature (as Jean-Paul Sartre's protagonist of his first novel, *Nausea*, Antoine Roquentine, reacted so dramatically to the bark of a chestnut tree). This juggling act is a physical fact that obeys the rules of mortality while striving to achieve permanence in a world of impermanence. The only logical resolution of that whimsy centers upon the nature of existence which, in turn, is indebted to nature, as continuously described by the ecological sciences in all of their cathartic emanations, none of which show up in the *Tractatus* because there can be no intellectual conclusion if all such summations are predicated upon lacunae, the essence of Wittgenstein's friend Bertrand Russell's famed Paradox.

This at once poses an insurmountable peculiarity – to want to be at the same time inside and outside the subject of our deliberations. We aspire to be the individual we suspect ourselves of being but want not to ascribe to such being, as if we inhabited an entirely different and inexpressible vantage point. The intellectual labyrinth comprising such a magnitude of differentials and dialectics is not easily explained. If a treatise can aspire in its accumulation of suggestions to be a mirror of ourselves, our private journeys, all of which will end in some form of death and anticipatory, cognitive abnegation and nihilism, then their concatenation, if it is to uphold some great truth, must nurture a precise point of clarity vulnerable to full elicitation, to a time and location of death, for example. We should be able to map mortality, produce the results of an autopsy, and give some guidance to future human beings who will themselves be caught up in the philosophical extremes of self-reflection and morbid denial, a far broader expansion of Al Gore's film, "Inconvenient Sequel."

But perpetually, the sole object of ultimate priority that hovers over Wittgenstein's posterity is the most uncomfortable of all mirrors, the one that pits philosophy against the individual, namely, that vast brilliant density of vagaries, Nature.

To speak of great truths is to assemble questions and answers with respect to the Wilderness, and to what we have called, elsewhere, the Other.

In attempting to clarify and convey methodical givens, line by line, for purposes of establishing a philosophical precedent, the beginning and end of time, space, of individuals, and the world, Wittgenstein's *Tractatus* is a surrogate for that ineffable compact that juxtaposes Individuals and Others on a world stage. It holds these two general categories – comprising living Beings – to be a specific and unique kind of promise made between our perception of the Cosmos and the fact that we live inside of it, inside, while attempting to see from the outside.

This is a feat to which philosophy has always aspired. Illimitable nuances, questions, solutions, adjustments, and variants aggravate the journey of mind as it enquires within the solipsist's boundaries. Whatever happens is essentially predictable. There are just so many neurons, cells, and atoms. Their interdependencies flourish according to any number of General Theories, such as Relativity. Equally exasperating, exceptions are multitudinous, just as patterns tend to elicit an equal number of newly discovered systems, regularities, and probability factors.

We cannot even approach the subject of coordinates on Earth. They are infinite, just as numbers are infinite. This would suggest that any and all grounds for a general principle cannot be accounted for in the long histories of human thought. They fail to surface even more rigorously in the thoughts of other species, for which no descriptive neurophysiology is qualified to make assertions. When we pose the question, for example, whether bees display happiness upon the acquisition of pollen, is that like suggesting a smile which accompanies a feeling of joy, or of gainful employment, or some other positive experience? We are grappling with words, arduously struggling like ingénues to attach bold certitudes – some kind of authoritative exclamation or scientific jargon to that which we cannot understand but are determined to label. Since the various periods characterized as "the Renaissance," we have increasingly taken it as an a priori truth that the Other, both as Individual and Collective, nurtures and hosts an exquisite infinity of musical presence, articulation, and Will. At one point this was determined to be the will of God; then of inchoate forces of the Creation; and then as random accidents or an infinity of prospects within a random universe that obsessed thinkers like Moses Maimonides in his *The Guide for the Perplexed* (circa 1190) or Thomas Aquinas' *Summa Theologica* (1265–1274).

That universe, for purposes of our own daily comprehension, is this world we cohabit, and it includes everything subject to dictation, as well as all that which is loosely described as ineffable. Wittgenstein's world is meant to be understood as *our* world and all worlds preceding this one and positioned to follow it. A book that is, in other words, a theory of everything, from the atom to the thought. The same could be said of other big books, by Hegel, Jean-Paul Sartre, Schopenhauer and the complete oeuvre of Picasso, Rembrandt, Shakespeare, and Mozart. The instrument of Wittgenstein's outpouring is neither poised, witty, nor easily digested because of its stark complexities that seek to apotheosize aspects of the living for which there is no *solace*, *security*, or *certainty*. These three nouns mean everything to our species but can never be handed over nor adequately conferred. There is no one to do so. Only God or its surrogates have the power over our imaginations to suggest inclusive force.

By embracing contradictory subject matter steeped in carefully orchestrated ambivalences, the story – not merely of Wittgenstein's *Tractatus* but of every sustained attempt to figure out just what is happening to us and who we are – pivots every assertion upon what is essentially an arbitrary precedent. We pirouette aboard a cumulative story line that offers to the intellect a prolific set of options, of more philosophy than we can absorb. All the while, we elevate our condition to that realm consisting of endless possibilities, as if to acknowledge our co-creative role in whatever lies ahead. This may be true or false. If it is only God that suffices to ensure the pertinacity of those aforementioned nouns – solace, security, and certainty – then we must acknowledge from the outset those same nouns' precise vulnerability; to Holocausts, the Death of Nature, nuclear Armageddon, and other globally debilitating anthropocenic consequences of our own behavior, as well as the tautological behavior of behavior.

Even if recent extinctions and other disasters were to be viewed as anomalies (the rate, pace, and escalations of which, over multimillion year background rates, they surely are), they cease to stand out when biology is juxtaposed with geology. In other words, when our evanescent time frame as a species is placed in a broader context, we recognize a double jeopardy that is historically fixed.

Hence, any reader is invited to invent arguments and interpretations that cannot fall outside of the contagion Wittgenstein has already predisposed us to. This is the essence of all normative cul-de-sacs. It is not a style so much as a finale, from the very page one of the *Tractatus*. A finality that carries in its ebb and flow all those who venture near to nurture syllogisms and vague equations as an act of homage to the book's rationale. These various efforts on the author's and reader's part together comprise humanity's attempt to address a futuristic, parallel type of thinker, of a personage presumably born into an altogether clearer, more logical universe. One who has the capacity to reject limitations of birth, genes, and history.

The Vienna Circle, in Wittgenstein's time, was frequented by many fine logicians, linguists, scientists, mathematicians, and historian of ideas. None could do much with the celestially sobering closure to which Wittgenstein was alerting them in the form of his *Tractatus*, his primary work. Additionally, he compiled a children's short dictionary, and his oeuvre includes the vast amalgam of Wittgenstein's *Philosophical Investigations*, published in 1953, 2 years after his death, which included his so-called *Blue and Brown Books*, lectures at Cambridge University during the years 1933 and 1935. In addition, as every student of Wittgenstein well knows, there were countless other posthumous publications consolidating approximately 20,000+ pages of notes, a life of thoughtful literary splinters and phenomenological provocations, both "sacred canon" and "scrapbook,"[7] many decades wandering across a veritable Sinai of aphorisms and dictums.

As two, man and woman, who have lived in and trekked throughout a myriad of Sinais throughout the world, we must wonder at statements like "the total reality is the world"; that "we make to ourselves pictures of facts," that "the picture is a fact," "a model of reality," and "linked with reality."[8] Such large categorical premises implode and leave us staring out a window with little to aid in our reflection upon things. As if that window were being swept by days and nights of rain, while we sat snug indoors listening, musing, nodding off. This is especially so at that moment

---

[7] See Gorlée DL (2012) Wittgenstein In Translation – Exploring Semiotic Signatures. De Gruyter Mouton, Walter de Gruyter GmbH & Co. KG, Berlin/Boston

[8] Tractatus Logico-Philosophicus, International Library of Psychology Philosophy and Scientific Method, by Ludwig Wittgenstein, With an Introduction by Bertrand Russell, Harcourt, Brace & Company, Inc., London: Keagan Paul, Trench, Trubner & Co., Ltd., 1922, p.39

Wittgenstein precisely envisions the following: "In order to discover whether the picture is true or false, we must compare it with reality"[9] and that "there is no picture which is a priori true "[10] and finally, "an a priori true thought would be one whose possibility guaranteed its truth."[11]

How does one *guarantee* a possibility? And who is this one, whose contingent reality hinges upon that "reality" itself? What is, after all, reality? Is a hummingbird's reality the same as that of a sea bass? Or that of the black bear who goes after the hummingbird's seed? Does a southwestern Haitian denizen with nothing left – nothing but her/his life itself or a refugee in Aleppo – after the latest catastrophes have everything that is human in common with a multibillionaire in her/his 30s, living in Silicon Valley, Moscow, or Shanghai?

That there is intense suffering at large is an indisputable truth. Ecosystems and individuals are finite. Our confident words and books belie our inability to predict anything. Control exists only to the extent that we live and die, show compassion or indifference, foster life, or kill her. Every thought and utterance is instinct with this do-or-die dialectic that has ensnared every possibility we might otherwise nurture.

By asserting the chaos of our lives throughout his seven propositions – facts, objects, logical atomism, etc. – Wittgenstein's famed conclusions are the benumbing *silence* that convenes upon the recognition of the Hegelian point counterpoints that are human history and the gulf potentially separating the individual from meaning, as some athlete from all athletics, to paraphrase the 60-year-old William Butler Yeats.

A Raison d'être

In this work we hope to inject a singular and palpable reality whose language manifestly lends philosophy a raison d'être. That language and thinking corresponds to all that is Nature, Biodiversity, and Eco-evolutionary Dynamics, as exemplified in one novice species, us. In contemplating the amalgam of biological events that have spun off *Homo sapiens* as one of her trial runs, the most critical of all questions must arise, if we are to survive: Is it incumbent upon us to try and be kind, gentle, and compassionate? Does the question hold only for individuals, or is it logical and inevitable that we pose this injunction at the level of the collective? Perhaps the odds of history and genetics are so stacked against any collective generalization as to devolve, all supposed ethics collapsing back unto the fragile locus of the lone individual, which, we would argue, randomly taps the tuning fork that may well be the fate of our species.

If this latter morbidity turned potentially sublime is the case, then one ponderous, enormous question looms large indeed: Are we still –in spite of every naysayer –

---

[9] Ibid., p.43

[10] Ibid, p.43

[11] Ibid., p.45

better off following the edicts of our hearts in the face of mass denial and the overturning of common sense by the majority of our fellow humans? Or are pessimistic adages the end result of every balancing act? If so, this would mean that our most self-defensive reflexes were our final fallback. And if such finality should test proof positive, then applied ethics, failing a critical mass among other humans, can only dissipate as we grapple with or strive to measure and comprehend the full extent of our incompetency as a species and nihilistic but *potential* state as individuals.

Histories of civilizations have collectively leveled enormous judgment upon the teleology of nullity in the wake of so much loss of faith, despair, collapse, and self-destruction. From Herodotus to Spengler, from Edward Gibbon, Will and Ariel Durant, Arnold Toynbee, Clarence Glacken, Paul Kennedy, and John Keegan to Irenäus Eibl-Eibesfeldt and Jared Diamond, there is an underlying philosophical disposition that has argued on either side of the individual and her/his ability to alter circumstances even minutely. Thomas Malthus and Paul and Anne Ehrlich have certainly chimed in on these weighing questions. But the teleological approach necessarily raises the notion of purpose and design, which Darwin and Wallace tried haplessly to reconcile with brute biological evidence. A divine Cosmos may or may not be a plausible mechanism beside any theory of evolution, although many, like Pope Francis, have sought to find a balance where science and religion are not incompatible.

Demonstrable proof of the individual's status amid a biological proliferation and ecological flux across the planet is simply unaccountable. The scientific community continues to debate the total number of suspected species with learned opinions often separated by utterly consequential powers of ten. Even if life forms, human included, could shoulder some consistency beyond their number of alleles or types of display, of shyness and boldness, of utterances and other songs throughout that haggard biogeography, we still are left bereft of the answers needed to solve the basic riddle, by itself, of human capacity, neural infrastructure, intuitive design, and our least accommodating abilities to grasp the very essentials of what is happening to us.

There are many who have argued that our cultural memes, as ethologist Richard Dawkins named it, have empowered the transmission of facts, fantasies, inclinations, norms, self-justifications, and any number of other impulse/reflex/purpose/intellectualization-driven mind-sets. These sets have proved both helpful and disastrous to our survival, as if survival on its face were everything worth talking about.

Our species' insatiable distribution is more easily tracked – not by metaphysics or philosophy – but according to metrics and flow charts, best applied in mechanical engineering, chemistry, astrogeophysics, thermodynamics, and other mathematical settings which translate into tools for calculating probabilities, swarms, events, and trends. Where religion and ethics lend a modicum of dignity to the individual, even some kind of promise, science is more likely to transform any individual into a host for yet other individuals, like bacteria in our gut. The miniaturization keeps going down the food chain into submicron-sized organisms and single cells, until our academic disciplines and, indeed, self-reflection, lose most traction. Between life and death – typically a short span – there is never a single conversation between the entities inhabiting one another, the nut to the tree, the beaver to its lodge, the millions of follicle mites living out their lives in our eyelashes.

Anthrozoologists and poets of numerous persuasion know that there is an ongoing dialogue in the world at large between creatures. But our rage for proof – peer-reviewed proof – has revealed a window on one great flaw in human nature: a kind of deaf and adamant cognition and corresponding cynicism.

In this work it is our hope to establish a new baseline for what a future species that merited the description loosely referred to as *humanity* might be like and what kinds of conversations it might have. We seek to pull the anchor from the muddied waters of a Wittgenstein's solitary crusade in the murkiness of illogic.

Some will recoil at the notion that this effort can only rely upon realms of zoological and evolutionary fantasy; that there is no purchase or verisimilitude that can be achieved by such speculation. We don't entirely disagree. But we also hold more than a little hope that wherever the zoological annals – caught in tempestuous winds of change – may be leading our species, we are more than mere hapless nomads condemned to our own shortsighted whims. Yes, Nature has always performed as she will. But it is clear to our consciousness that she is never neutral, even if science insists she is. Swarms of earthquakes in Oklahoma, allegedly as a result of regional fracking, pose one example of this. Is it possible for individuals or the human collective to favorably sway nature's predilections? We have certainly forged more than a few diversions, every hydroelectric plant, stem cell, or barbed wire fence being but a few examples.

While *H. sapiens* have strenuously sought to reconcile Earth's apparent priorities with our ever expanding perception of human hegemony, even since well before the dawn of what is widely regarded as science, we find that we simply cannot in good faith condone those many Armageddon-driven prospects that have long been the rage of our species' entertainment. The two most successful films ever distributed were disaster movies ("Titanic" and "Avatar"). More to the point, the twenty-first century, as we conventionally refer to this current phase of the Anthropocene, an epoch of biological ruination across the planet solely unleashed by humans and registered in geological time, is rapidly undoing long annals of Earth's efforts to establish a philanthropy of life forms, as witnessed at the approximately 3.7 million-year-old endolithic lichens and stromatolites at Shark Bay, Australia and the Isua Greenstone Belt (Archean) in southwestern Greenland, dating to nearly 3.8 billion years.

We first hope to establish the nature of that conflict by questioning the basis for humanity's self-aggrandizing principles. Two famed statements come to mind in rendering this query that would undermine any human-related balance. First those words of Macbeth, "…why do I yield to that suggestion, whose horrid image doth unfix my hair, and make my seated heart knock at my ribs/against the use of nature?"[12] And that remarkable statement by Francis Bacon, at the near conclusion of his *Natural History*, "…I do not understand, but that in the practical part of knowledge, much will be left to experience and probation, whereunto indication cannot so fully reach; and this not only in specie, but in individuo [sic]."[13] Inconstancy

---

[12] Shakespeare W The "Tragedy of Macbeth," Act 1, Scene V, 1605–1606

[13] The works of Francis Bacon, lord chancellor of England, A New Edition: With a life of the author, By Basil Montagu, Esq, in Three Volumes, Vol. II., Philadelphia: A. Hart, Late Carey & Hart, 1852, p 136

of behavior and ideal is unleashed upon our world by man, with so unsettling a conclusion as to demand a common sense tempered still by love.

In her fascinating 1998 tome, *Unnatural Selection: The Promise And The Power Of Human Gene Research*, Lois Wingerson charted the medical purview of that inconstancy with respect to genetic research and the counseling of patients by specialists rightly nervous about their power over the individual, against a backdrop of medical ethics that were turning topsy-turvy. In her chapter uncannily titled "Identity Crisis," Wingerson described how primary care doctors were having to reexamine (if not entirely reinvent) their own agendas, perspectives, and bedside linguistics in the face of what was veritably a new genetic lexicological avalanche. "All of this takes time. In the near term, we may be left truly on our own. We may be offered testing we don't want or denied testing we do want. We may have helpful counseling, unhelpful counseling, or no counseling whatsoever. As things stand, it's anybody's guess."[14]

We hope, then, in confirmation with the above premise of tempering our conquest of nature in an ill-advised attempt to hail the Self, to question individual and collective mechanisms of change that human nature has consistently but not entirely embraced. The norms but also the exceptions are of equal interest to us and span speculations on any number of human agencies: genetic, moral, political, civic, legal, socioeconomic, ethnographic, and particularly those born of aesthetic predilections and natural sciences.

Finally, we want freely, and by invoking the broadest exercise of first principles and deductions, to dream of a newborn in the name of humanity whose time has come, surpassing what is otherwise a morgue of largely embittered Darwinian-projected memories, an ecology of cemeteries.

There have been societies and tribes, small groups like the Toda and Tasaday, and others who throughout history have provided windows on humanity in ways typically overlooked by the philosophical and ecological communities today. Small, isolated hamlets that have confounded the near mathematical certainty of a species destined to fail, a collective of hominids on every continent that is in a mode of ecological self-destruction with seemingly no end to the misery in sight, a reality accented by the August 7, 2017, combined release by 13 US Federal Agencies of the severely escalating climate change impacts across every continent.[15]

But there have been sociological counterintuitive success stories among our kind. One such confederation of peoples once inhabited what is northeastern Mexico. They may stir echoes of Mexico's 50,000 or so remaining Tarahumara (Rarámuri) or fewer than 700 Lacondon-speaking peoples. The tribes of Mexico's region (State) of Tamaulipas may have been unique and are the subject of the following chapter. Who they were may hold enormous promise for what we may yet become.

---

[14] Wingerson L (1998) Unnatural selection: the promise and the power of human gene research, Bantam Books, New York, NY, p 262

[15] https://www.nytimes.com/2017/08/07/climate/climate-change-drastic-warming-trump.html. Accessed 7 Aug 2017

Chapter 2
The Lost Tribes of Tamaulipas

A Primordial Consciousness in Northern Mexico

There exists a place both on and off the maps. Contiguous with the 144,530 hectare Reserva de la Biósfera El Cielo in the Sierra Madre Oriental, it is the most northerly, non-contiguous tentacle (surprisingly) of Amazonia habitat, with a high density of resident jaguars. Therein lies a series of lushly vegetated mesic transmontane valleys, peaks, and gullies gorgeously spread out in a virtual cartographic anonymity. Where one finds only the rare farmer and a few pitted dirt roads that at times defy the most robust four-wheel-drive vehicles, between the Sierra de San Carlos and the villages of San Nicholás, in what today is Tamaulipas state, in north-central Mexico. The area in question lies a mere 150 miles southwest of Brownsville, Texas, and is known as the Burgos Municipality. Never have two different worlds – South Texas and northeastern Mexico – been so physically close and historically removed, phantasms of enormous ecological and ethnographic significance.

It is in this Tamaulipas state, famed for its archaeological museum in Tampico, that we managed to explore most of the 11 recently discovered rock-painting sites – walls and caves – containing over 5000 works of art, the majority of which, pictographs, were first surveyed in 2006.[1] However, one of the rock walls had been found by a local farmer just days before our arrival, high up in a forest where an ominous sign had been left nailed to a tree down below, a 20-foot dead boa constrictor. The result, perhaps, of drug cartels which use these off-the-map gullies to run some of their product northeast from the state of Sinaloa. Mexican paramilitary presence, throughout both Tamaulipas and neighboring Nueva Leon, corroborates this treacherous reality. But other than that farmer and his dog, we were probably the only others to have stood before this wall in as many as 4000 years; some suggest 8000.

---
[1] http://www.npr.org/templates/story/story.php?storyId=208543543; See also http://marginalia.lareviewofbooks.org/gustavo-a-ramirez-on-the-rock-art-of-burgos-tamaulipas-mexico/. Accessed 15 Oct 2016

Indeed, many were now calling the complex of pictographs and petroglyphs as Mexico's Lascaux.[2]

It took 2 years to get there, given the geopolitical turbulence. Thanks to our colleague Gustavo A. Ramirez Castilla, then director of the Mexican Network of Archaeology ["RMA"] and researcher of the National Institute of Anthropology and History of Mexico in Tamaulipas state ["INAH"], as well as chief curator of the museum in Tampico, and with his trusted and long-cultivated contacts in Burgos, including the mayor, we were finally able to concretize our field research journey.

The possibly three different groups of people who painted this abundance of deeply alluring work were completely unknown to Mexican archaeologists well in to the twenty-first century. Those peoples left no objects, none yet discovered – no musical instruments, dyed fabrics, pots, agricultural, or hunting materials. Nor does any evidence purport to violence, skirmishes, or one tribe hunting, or competing with, another; there is no record of pursuits. And it is more than likely that no Spaniard knew of these enigmatic denizens, ghosts we had come to study. Their social quintessence incites pragmatic and metaphysical ponderations that left us – on site – both astonished and humbled. We are the same personages, thousands of years later, standing over the sixth extinction spasm and gazing back at the first of our kind to have entertained violence or non-violence, ecological pacifism, or selfless indifference. Did they harbor these sorts of deliberations? Did they have choices predicated upon dire dilemmas? What were they thinking? How much have the habitats and species abundance changed since they occupied these caves, passed along these riparian habitats, dreamt of some horizon? Save for the spectacular body of artwork that leads present speculation toward a convergence of queries, we have no idea who they were. Or how many of them thrived in these wildlands. One thing is certain: a votive and ebullient individualism exudes wonderment wherever they chose to leave something behind and before.

We found ourselves, day after day, sitting and standing beneath painted cave walls discussing all of these enigmas – human dimensions of historic, philosophical, environmental, and archaeological significance germane to our humanity today. These lost tribes of Tamaulipas speak to the eternal. We enter the conversation knowing that, today, we reimagine the inception of these spectacular depictions – but of what? And why do we need to know – beneath an enormous burden. At stake is the very neural infrastructure, that human capacity to speak to who we are and the complexity of our species' interconnections to the Others. Or that's the archae-logic at work, our divining rod. Following our week-long expedition to the many varied sites, all on gorges and upon mountains requiring arduous treks and, in one case, masks to protect against possible airborne diseases spread by some unknown

---

[2] Megan Garber, Discovered: a cave art complex that could be the Lascaux of Mexico – the paintings feature humans and lizards and centipedes, and could be 8000 years old, http://www.theatlantic.com/international/archive/2013/05/discovered-a-cave-art-complex-that-could-be-the-lascaux-of-mexico/276298/. Accessed 15 Oct 2016; see also George Dvorsky, Thousands of cave paintings have been discovered in Mexico, http://io9.gizmodo.com/thousands-of-cave-paintings-have-been-discovered-in-mex-509994795/ Accessed 15 Oct 2016; See also, http://www.bbc.com/news/world-latin-america-22632301/. Accessed 15 Oct 2016

pathogens, we arrived at one definitive conclusion: The context of North American history at the time these paintings were executed, and in light of our ecological crisis today, does not provide for much illumination, though great personal pleasure. As has been the case with so much primeval rock art throughout the world, we simply do not know what it means. Instead, we have to *feel* our way back in time and place to put ourselves there.[3]

Jane and I are indebted to Dr. Castilla for his most generous sharing of his own archaeological and emotional insights with us, both on site throughout the topography of this painted world across Tamaulipas during mid-October 2013 and in the years since.

It was a vast "unhappy drama," as Castilla calls the 250-year period of Spanish domination of Costa del Seno Mexicano/New Spain's New Santander, today known as Tamaulipas. But that relative modernity enshrouds the many worlds of those Indians who had previously inhabited one of the most biodiverse regions in North America. They lived here long before the arrival of aggressors bent upon the land's colonization. Writes Castilla, "With the advent of bow and arrow technology around 500 A.D., Indian life in the region changed radically. But it did so again in 1748, when an army of 750 soldiers commanded by two men, Colonel Jose de Escandon and Helguera, Count of Sierra Gorda, arrived at Seno Mexicano with a caravan of settlers; families of ordinary people, mainly mestizo, Creole and Christianized Indians as well as a small number of other fortune-hunting peninsular Spaniards who were new to this essentially Wild West…." And thus begins a most depressing history of deftly planned conquest, attempts at religious conversion, and ultimately, concerted efforts toward extermination of the indigenes who were resisting religious conversion.

Linguistic evidence suggests there might have been many dozens of indigenous Indian populations in Tamaulipas prior to the arrival of the Spanish. In fact, in 1945, Gabriel Saldivar referenced more than one hundred such groups, and the Spanish Franciscan priest Father Vicente de Santa Maria (1742–1806) had noted various customs of local Indians. Yet, as recently as the early 1800s, Castilla believes there were less than one thousand local Chichimeca Indians and, today, "not a single known native Indian in the whole of Tamaulipas." There has even been an "extermination of ancestral memories," he adds.

Early Spanish Castilian designations for some of the encounters with Indians, some of the words indicative of reality, others simple pejorative, suggest not only

---

[3] See Dr. Castilla's 45-page as yet unpublished essay, "Las manifestaciones gráfico-rupestres de Tamaulipas" (2014), which we translated from Spanish and comprises the fullest contextualization for these cave art sites, to date, along with a major, previously completed thesis by Dr. Castilla's colleague and friend, INAH researcher Martha García Sánchez, "La presencia del arte rupestre en Burgos, Tamaulipas," Tesis de licenciatura, Unidad Académica de Antropología, Universidad Autónoma de Zacatecas, Zacatecas, 2012; and the largely pictographic brief by Jorge Luis Berdeja, "Una galería de pintura rupestre en la Sierra de San Carlos Arqueólogos del INAH documentaron las obras realizadas, al menos, por cinco grupos de cazadores-recolectores nómadas" http://www.inah.gob.mx/images/boletines/reportajes/20130919_pinturarupestre/rupestre.pdf. Accessed 15 Oct 2016

various rituals of mourning, a variety of customs, and tools like metate grinding bowls but also "local indigenous habits of consuming uncooked foods." [The word] "scratched might mean to have been covered in body painted lines, or painted in the image of black teeth around a person's mouth," for example. Castilla points out that some names were lent "to the nomadic tribes of northern Mexico by the Aztec Indians who accompanied the Spanish conquistadors to Zacatecas. Chichimecatl [for example] is a pejorative Nahuatl word meaning 'those who speak like dogs' or 'lineage of dogs', meaning it (he/she) did not speak as individuals, they were animals or wild." Generally speaking, Chichimecatl came to denote indigenous barbarians, a racial slur that probably continued for several thousand years, certainly during the late eighteenth century when one Gaspar de Portola, at the request of King Charles III of Spain, sought, with his confederacy of aggressive Franciscan "missionaries, to convert between 133,000 to more than 700,000 Native Americans – representing more than 100 tribes" to Christianity, probably commencing in 1769. These included the Kumeyaay (Tipai Ipai) in Baja and what is today San Diego County and another, unknown tribe that flocked with the tools of their gorgeous pictographs to what is the last remaining relatively pristine land grant in San Diego County, Rancho Guejito, and one massive boulder in particular – where ferns and trees and other images were painted, probably while fleeing from the Franciscans: nearly 5000 years after the tribes in question in Tamaulipas.[4]

The history of conquest of North America is, of course, rife with a blurring of meaning between human individuals and animal individuals. It is a disastrous motif that crosses all symbolism, gut instinct, and reality.

Despite intense linguistic research to help revivify a better understanding of the vanished Indian tribes, with work in such groups as the Uto-Aztec and Hokan-Coahuiltecan, only those Huastecos still living in the southernmost corner of Tamaulipas and part of the greater Mesoamerican cultural network offer even the most remote possible cultural and/or ethnic connections to the people that once lived farther north, apparent hunter-gatherers with a variety of well-recorded survival modes, what future generations simplistically think of as customs or rituals or worldviews.

What is clear, however, is that "the first inhabitants of Tamaulipas arrived about 12, 000 years ago," Castilla and colleagues have suggested, despite the physical evidence to date suggesting no older remnants than 4000 years, a factor of weather and deterioration. Some of those remnants include parts of, or entire mummies like

---

[4] Serrano, Carlos Mireya Montiel, Gustavo A. Ramirez Castilla (2008) Osteological analysis of a funeral in northern Tamaulipas Late Archaic (4500 AP). International Symposium of Physical Anthropology Juan Comas, pp 61–62. See Department of the Interior, Bureau of Indian Affairs Notice 145A2100DD/A0T500000.000000/AAK3000000: Indian Entities Recognized and Eligible to Receive Services from the United States Bureau of Indian Affairs. Federal Register, January 2015 (PDF). Federal Register. 80. Government Publishing Office. January 14, 2015. pp 1942–1948. OCLC 1768512. See also Harry Jones J, Pictographs of a dramatic past. *Los Angeles Times*, Dec 25, 2015, p B3

the famed "Woman of Corindón" and burials and mummifications studied at Little Hell Canyon, the Enchanted Cave, and elsewhere.[5]

While the first three Tamaulipas cave art sites all in one canyon were explored in 2006 by INAH researchers at the request of the city chronicler of Burgos, Mrs. Cecilia Homes, additional research was soon mounted in subsequent years. By late 2013, with at least 11 known sites and a huge probability that scores of other caves and cliffs remained to be discovered throughout the labyrinth of canyons crisscrossing the Sierra de San Carlos, it became clear to all of us who had had the rare privilege of climbing to these locations, that here was undoubtedly one of the most important UNESCO World Heritage candidates in all of Mexico, certainly on par with the UNESCO-recognized Algerian paintings of Tassili n'Ajjer or Spain's Altamira.

At the northern epicenter of this aesthetic fabric of prehistory is Burgos, a splendid little town out of time, founded in 1749 and devoted to Our Lady of Loreto, in whose name the church in the town plaza was completed in 1792. The great enigma hovering over this village before time is its astonishing proximity to the United States. With some 2000 residents, there is virtually no tourism to this region and just one, very lovely B&B run by a local couple. The drug cartels have all but destroyed access to Tamaulipas. Other forces of modernity, however, are piling up in the guise of the Burgos Basin natural gas boom. As Castilla notes, "Roads are being built, along with gas stations and the arrival of outsiders. There will be serious socioeconomic and cultural fall-out to the region, with no plans in place, at least as yet, for mitigation of negative cultural or ecological impacts."[6] At those gas stations and patrolling the roads to fight drug traffickers are military police which does not exactly invite outsiders into these various ejidos or traditional communal properties that make the geographical map of the region particularly complex.

All this, in vast contrast to much of Mexico's other archaeological wonders, not to mention those petroglyph-rich regions throughout the United States where enormous amounts of tourism, national park and monument legislation, and in situ Native American tribal police and communities have worked to combat the escalating crisis of vandalism.

It must be noted from the outset, however, that as yet there appears to be no known connection or connectivity in terms of artistic styles and, most probably, among the peoples themselves who committed their lives to this unique habitat in

---

[5] For extensive bibliographical materials on the mummies of Tamaulipas and surrounding regions, see citations in Castilla (2014) including Josefina Mansilla Lory and Ilan Leboreiro Santiago Reyna, "2009 Report on the Mummies of Mexico Project DAF/INAH, on the advice, analysis and study of the mummy found in the Hidden Canyon, municipality of Llera, Tamaulipas, in collaboration with INAH, Tamaulipas, Mexico. Center Technical file Tamaulipas INAH Center; MacNeish, Richard S., Excavation in the 1998 Preliminary Ocampo Region of Tamaulipas, Mexico, Archives of the Andover Foundation for Archaeological Research (AFAR), Peabody Museum. Andover, MA. (Unpublished); Ramirez Castilla, Gustavo A. (2014) Funeral Traditions, Premature Burials and Mummification: Advances in the Mummies of Tamaulipas Bioarchaeological Project. Yearbook of Mummy Studies, vol 2: pp 133–142

[6] Op. cit., Castilla 2014

northeastern Mexico thousands of years ago. We believe these tribes were, if nomadic, only partially so, and likely remained within the general areas of their paintings, that region between the tropical Sierra Madre Oriental and thorn- and scrub-dominated Sierra Chiquita, the 2320 square kilometers of limestone-layered, oak- and pine-forested cordillera with its unique volcanic tepuis, across Tamaulipas, heading into the southern state of Nuevo Leon. Such botanical expanse expresses the northern limit of the cloud or fog forest, with high species endemism, unique to the biologically waning 36 terrestrial hotspots in the world.

Because there are numerous and easily accessed GIS portals in which the various cave site denominations have already appeared, we are going to forego naming the 11 canyons and caves in the interest, to the extent possible, of discouraging future visitation. This is both ethical and practical. There has already been some minor vandalism of pictograms by locals, and our intent is not to elaborate upon the thousands of images, but, rather, describe only a couple of them, rendering various details and suppositions, emphasizing that these are purely speculations. We really don't know what they mean to depict. The species, the gender of humans, the overviews, and the spiritual intentions – of which there are partially evident deep structures and vast philosophical menageries – are all open to wondrous suggestion, even in comparison with other similarly dated or earlier paintings at a host of sites on most other continents.

What we will say is that these paintings are mostly on limestone, sheltered within canyon overhangs, typically high above, or nearby extensive and colorful gravel beds during the non-rainy season. As Castilla has elaborately detailed in a brief summary of each of the primary 11 known sites to date, the paintings vary from bichromatic to polychrome (multiple colors) emphasizing, typically, a broad range: "…of shades of red, followed by black, yellow, ocher and white….There are geometric forms, anthropomorphic, phytomorphic, zoomorphic and a large group of unidentified forms of expression... The anthropomorphic features include a variety of figures full-bodied, with or absent a head, with or without upper extremities. Sometimes the sex of an individual is distinguished; sometimes there is touching, or an actual ornament fitted upon a head. Note that in some, there are only three fingers in a hand. Some of the figures appear alone. Others are participating in scenes that seem to represent rituals. Usually, the forms are very basic, with dots and circles. In the case of shapes reminiscent of branches, leaves or plants, thus far their species identification has proved all but impossible. In the case of the geometric images, we see clearly delineated straight lines, curves, zigzags, intersections, parallels, angles and circles, diamonds, grids, rectangles, etc."[7]

There are stars and meteorites, perhaps a sun, a moon. And countless species. Castilla believes most were painted by hand – one, two, or three fingers. Some, possibly, by painterly novitiates. But we're not sure. An alleged depiction of a centipede might just as easily be a maggot, a moon, a fetal position, or half-sun, this latter comporting with some of the mummification positions reported elsewhere in Mexico. And so it goes. Questions. But what is unambiguous are the following

---

[7] Ibid., Castilla 2014

characteristics across the full spectrum of these paintings: a primordial consciousness unique to the evanescent peoples who populated this corner of the world. At one of the caves, Castilla believes, we can decipher: "cactus, shrubs, leaves, trees, sometimes zigzagging lines "suggestive of" magueyes (Genus Agave, e.g., mescal) which might in fact symbolize water, as well. There are anthropomorphic characters representing both full, sometimes abstract sometimes naturalistic figures; body parts, protruding arms, heads and a potential phallus. There appears to be some kind of headdress on one of the figures. Additionally, there are lizards, millipedes and other unidentified and repeated creatures."[8]

Amid the cosmic proportions of, possibly, "astronomical figures" and designs suddenly appears a procession, tepees, possibly a boat, stairways to heaven? Were these creeks and rivers much wider 5000 years ago?

Then comes the contradiction: driving 30 arduous miles on what barely constitutes a road and then trekking a mile high into the forest, we come upon rock images of atlatls – spear-throwing devices like those used in ancient Greece; and darts, at first glance. Perhaps shields not unlike those one would have seen wielded during the Peloponnesian War. There are "group meetings" and unisons. The paintings all seem oriented geographically. A strategic river? A convergence of cliffs? A safe place, a not-so-safe spot within the map of their world? And the posthumously received notion that there was not one, or two, but likely three different groups who committed their handprints to these rock surfaces, a notion predicated upon stylistic differences and similarities over the hundreds of square miles Castilla and his colleagues have intensively studied.

Within the canyons, and beneath the cliffs, we found an abundance of arrowheads, flakes of stone no doubt used for the fabrication of tools, for rendering incisions, sharpening, and fashioning. Our standard of confrontation was of a strict, short ethical leash – disturb nothing, remove nothing. Some of the locals who have known about a few of these marmoreal concatenations of aesthetic purpose and reverie for at least three decades have been less discrete in their fascination, collecting hundreds of ground-based artifacts and keeping them arrayed on tables at their homes nearby. One can hardly find fault with the fascination that arises at the intersection of the twenty-first century and a 4000–8000-year-old Weltanschauung made manifest all around one's deeply isolated contemporary dwelling places. This urging to be connected to a geographically fixed and local past is tantamount, it would seem, to being human. Building for 199 years a Leaning Tower in Pisa, for example, knowing, hoping that one's descendants would feel enthralment beneath the quirky engineering. Such hopes are never relinquished in Mexico, despite many troubled séances. As early as 1910 in his spectacular work, *The Ruins of Mexico*[9] looking at Chiapas, Yucatán, Tabasco, Oaxaca, and Puebla,[10] Constantine George Richards had written:

---

[8] Ibid., Castilla 2014

[9] Chiapas, Yucatan, Tabasco, Oaxaca, and Puebla (1910), In: The ruins of Mexico, vol 1. H. E. Shrimpton, London

[10] Ibid., vol 1, H. E. Shrimpton, London, pp v–vi

"There is hardly a town or village that has not got what they call their 'Pueblo Viejo,' or old village near by (sic), generally situated on a hill or some high point, which show that Mexico in ancient times as thickly populated...in many of them still exist the primitive walls with hieroglyphics and the basements of temples, terraces, forts and other buildings. Some are known to belong to certain races which inhabited these regions, but most of them belong to races unknown to-day, and when the Spaniards came to conquer Mexico they found them in a similar condition...".

When we scan the vast aggregates of such ruins, like those at Uxmal or Chichen Itzá, both in the Yucatán, or the ruins of Mitla or Xoxo (the "Toad's Cave") in Oaxaca, we see a Mexico teeming with a richness of life histories. Every biology translates into a biography. Historiography, abetted by such hieroglyphs, can resurrect every lesson we are prepared to read into our ancestors. But when cultures arise and disappear amid a pantheon of mysteries, our subjectivity freefalls. Whatever path-finding methodologies we should hope to apply in an effort to make surefooted inductions are simply a lost cause. We are on our own. But this is also wonderfully liberating, the idea that we may not, after all, be doomed to repeat our histories, whether the confused and often imploding transpositions of Whig and Tory factions, for example, of gruesome civil wars and ill-conceived crusades, or of rapacious turns and stupid bombs, one madness or collapsed civilization atop another, and now, the cumulative ecological effects of all our internecine politics and economics, industry, greed and indifference, the Anthropocene.

In ancient Tamaulipas culture, we are privy to poignant modernities and conjectures about the meaning of life. There is excitement and insight preserved on stone that gives us some idea of what those of us alive in the near future *could* look forward to. What we have elsewhere termed *the hypothetical species.*

Nearly every well-studied site of prehistoric rock art throughout the world suggests a transfixed consciousness that we tap into the moment we meditate upon it. Particularly when we do so on site. Being there, where ancestors stood painting their hearts out, acutely observing the world around them, taps a quality of perdurable exhilaration that knows no bounds. It is art at the essence and her urgings plumb that genetic fixity somehow wrapped enticingly and forever undetermined around a vortex of the clandestine and covert anomalies of human life that all great art excites.

Such modalities express a lasting hope in the guise of a human hand, of a bison or a ladder. These tender, if tenacious, nodes of consciousness seem devoted to interdependencies that stem from perception beyond the ordinary. This is key: windows on other worlds, eyes of intelligence gazing through those windows. What we cannot say with any confidence is whether those worlds are within us, before us, or beyond us. What we *can* intone is that humanity has, for tens of thousands of years, expressed the reality of biospheric love and compassion, of biophilia as determined colorfully and artfully on tactically (possibly strategically) chosen walls of rock, from Australia to Spain and France, from Russia to Argentina.

In the United States, we might well speculate on contemporaries of this Tamaulipas consciousness, suggesting an ecologically sustainable collective whose precolonial collectives and the perceptions of all those denizens arguably advance a crucial vision of what human life was and is capable of in the future. But we also know that that extraordinary diversity of tribes and languages, of geographical adaptation and a through-story of palpable relationships to the natural world, was overrun by invaders – as in the case of northern Mexico – other humans who had two primary motives: greed or flight from bondage. In both matters, the narrative involving tens of thousands of years of North American habitation quickly went from what we might adduce as a romance to terror and misery, transforming the pastoral into one of the vast tragedies of human history.

While many noble poets, chiefs, shamans, and unheralded indigenous families stand out as remarkable torchbearers of truth and insight, it is important to recognize a single American photographer, Edward Curtis, who came to champion both that imagined romance but also the reality of what he called "a vanishing race," in hopes that posterity would remember these people and, just possibly, revivify their legacy in our own children and grandchildren to come.

Chapter 3
Edward Curtis' Vision of Transcendence

North American Ecological History Leading to Edward Curtis

The celebrated passions of Edward Curtis did not arise inexplicably. Their trenchant origins – a spiritual orientation to the natural world and all those who dwell therein – invoke those most elegant of human responses to the environment: compassion and art. Both of these psychological and emotional "triggers" may be hardwired in our genes, something easier said. They are certainly keys to the survival of our species: our ability to celebrate and revere nature. And both qualities are to be found in a compelling biological history leading up to the pellucid and overwhelming love that Curtis felt for Native Americans, the ecosystems they revered and upon which they vividly depended.

In contemplating between 300,030 and 60,000 years of Native American ecological and cultural history (paleontological debates rage), we would argue that the emergence toward the end of this dazzling timeline of a humble Renaissance man named Edward Curtis can be seen to have written all through it a sense of gorgeous and critical destiny. Curtis' legacy of paying tribute to Indian sacred life, thought, and culture was no accident. And with sufficient hindsight – and the extensive documentation that has attended Curtis' extraordinary career, most seminally by scholar/artists like Christopher Cardozo – there is every reason to believe that Curtis was truly "meant to be," a critical aesthetic and psychological link helping to guarantee the survival and pertinacity of that indigenous humanity which early on came to inhabit North America, at no meager cost.

By that humanity, we are referring to those many discrete populations who first – according to the genetic and archaeological evidence – migrated south from Beringia all the way to Tierra del Fuego along the Pacific Coast, using innovative microblade technologies, canoes, fire, and inventive culinary appraisals of the natural world in to which they were immersed. These peoples were brilliant technicians, spiritual ecologists, and sure-footed pragmatists. Their art, customs, rituals, and orientations to their immediate surroundings read of a biological science mastered tens of thousands of years prior to their emergence on North American shores. Comparable deep humanities can be read into the paleontological records from the Zagros Mountains of modern-day Iraq to islands in today's eastern Indonesian archipelago to the central regions of the Amazon. A dispersion of seemingly like-minded communities engrossed by what might generally be characterized as a deeply spiritual, typically animistic local as well as a world view.

Mythologists and anthropologists like Mircea Eliade, Claude-Levi Strauss, Paul Radin, and James G. Frazer have variously assessed the personality, cartography, and ontology of these innumerable cultures and cultural perceptions, forming the basis for our received notions about the many cultural mother lodes that give us to better find ourselves. We are indeed the sum total of their many Bibles and legends. Some of these belief systems have endured, while others succumbed to the challenges of a frequently hostile modernity, or lay buried, "mute stones," as archaeologist Paul MacKendrick has likened them, hidden or obfuscated within the psyches that become who we are, or we cast off as time and circumstance permit or dictate.

Other unique cultures remain threatened in the guise of the last remaining "uncontacted" tribes, as well as hundreds of millions of indigenous peoples struggling for basic human rights in a twenty-first century largely indifferent to their *old-fashioned ways*, variously failing to recognize that those very "ways" are at the heart of the human experience and of our humanity, just as the "lost tribes of Tamaulipas" remind us individuals who comprise communities of the very meaning and critical worth of individuals to the integrity of ecosystems. That integrity is a pillar of the world, the only one we can possibly know and hope to function in.

Archaeological Evidence of the Cultural Vortex that So Inspired Edward Curtis

Toward the latter millennia of the aforementioned migrations, peoples like the Southern Miwoks and Paiutes, for example, settled in central and Eastern Yosemite, California, while others fanned eastward across North America and made this continent their home at a time when their own populations were stable. Glaciers were receding, and the introduction of varied fire-clearance approaches to creating habitat margins engendered rich and sustainable food supplies.[1]

In Yosemite alone, over 1500 Indian sites spanning 6000 years of documented habitation have thus far been uncovered by archaeologists. Because this comprises less than 10% of the park's 1200 square miles, it is clear that thousands of other sites remain to be discovered. The utilization and occupation of Yosemite by non-Indians has spanned approximately 168 years. The early invasions by outsiders required the presence of the US military to protect the habitat from direct exploitation, largely lumber extraction. In fact, the famous Wawona Tree, a giant sequoia 227 feet high,

---

[1] "References On The American Indian use of fire in ecosystems," compiled by Gerald W. Williams, Ph.D., Historical Analyst, USDA Forest Service, Washington, D.C., June 12, 2003

26 feet wide, collapsed during a winter storm in February 1969. It had been carved clean through for vehicular traffic more than four decades prior to its fall. Indeed, during the year 1923, the one-millionth automobile was driven through that hole, bearing a symbolic clamor of Americans who had been promised by this nation's first national park director, Stephen Mather, various scenic wonders and outdoor adventures all to be delivered on a kind of silver platter and in a style of comfort to which modern America had become accustomed. That was the same year the last full-blooded Miwok resident of Yosemite was essentially extirpated from the sacred valley.[2]

At Yellowstone, America's and the world's first national park, consecrated by President Ulysses S. Grant in 1872, a rash of inconsistencies and outright humiliations of a sort consistent with all those other interactions between indigenous and nonindigenous aspirants on a continental stage, were played out. Yellowstone's Indian residents – Bitterroot Salish (Flathead), Crow, Shoshone, Bannock, Cheyenne, and Nez Perce – had lived in and migrated back and forth through Yellowstone for at least 11,000 years. Obsidian tools fashioned in Yellowstone would be traded all the way to Illinois with other tribes. But when Yellowstone National Park was consecrated – Congress determining there was nothing of mineral worth there to be exploited – not a single Indian was consulted.

When one of the earliest tourist groups from Montana happened upon some Nez Perce residents, the park's public relations people thought the encounter nightmarish, characterizing the Indians as "wild renegades" who did not belong there.[3]

The situation would go from bad to worse, as subsequent park authorities accused the Indians outside the park of creating fires, of poaching, of scaring away paying tourists. Any so-called conservation ethic in the hands of the actual land managers was a bogus and tragic continuation of a disastrous script, one more step toward a near total ethnic cleansing well underway throughout the nineteenth century.

It was not until 1961 that the US National Park Service issued a statement conceding that "Evidence collected over the past several years seems to indicate that many tribes have been more or less permanent residents of this geologically mysterious area." That residential pertinicacity coincided with a greater ecosystem of nearly 19 million acres inscribed by the most active thermal machinations and other marvels on the continent, as every park enthusiast well knows, and a most multitudinous vertebrate, the bison, a fantastic creature whose lives and deaths were vastly intertwined with the Plains Indians and which, not unlike the indigenous peoples themselves, was nearly driven to extinction, a story surrounded by multiple windows of tainted glass.

The northern plains-roaming ultimate ungulate, the even-toed bison, numbered, even into the early 1800s, over 60 million. By 1833, the animal was extinct east of the Mississippi River and, by the late 1800s, numbered far fewer than 100 individuals across all of America, most of those survivors hiding out amid the steam of the

---

[2] Archaeologists at work – Yosemite National Park. www.nps.gov/yose/historyculture/archeologists.htm. Accessed 20 May 2015

[3] Ojibwa National Parks & American Indians: Yellowstone. http://nativeamericannetroots.net/diary/688. Accessed 21 May 2015

thermal vents of Yellowstone. Their fate came a breath away from that which drove the passenger pigeon to actual extinction, a mindless syndrome of callous stupidity on the part of humans that connotes an all-out assault by the majority, in political terms, on a minority. Ecology and politics intersect with unholy, archaeological clarity, time and again.

Amazingly, park service and state wildlife managers surrounding Yellowstone continue to insist on the mass killing of the recovering species that has always been perceived as sacred to every Plains Indian tribe, not that Indians themselves were not frequently involved in the killing of bison. But in 1997, Yellowstone National Park officials sanctioned the slaughter of 1100 buffalo under the pretext of an institutionalized assumption that these animals inevitably transmitted *Brucella abortus* to domestic cattle, animals doomed to be slaughtered, in any case, for human consumption. But that assumption – as with scrapie and prions – has long been scientifically questioned.[4]

Mixed Affairs

In contemplating so many mixed affairs, it is important to realize that Paleo-Indian migrations tens of thousands of years ago began under conditions of environmental stress: survival, even in a veritable garden of demographically relative Eden, was never predictable. Any doubts are easily acquitted upon scrutiny of the mortality rates stalking the first European colonists to America's eastern seaboard. The three ships with British settlers arriving at Chesapeake Bay in the spring of 1607, there establishing the Virginia Colony – 13 years before the next wave of pilgrims at Plymouth Rock, Massachusetts – enjoyed no springtime. Two-thirds of their legions perished from disease and starvation.

The English survival skills hit upon hard rock, frozen winters, and a severe learning curve in terms of food production, while gold and tobacco were also foremost on their minds (and the coveted taxes sent back to the British Crown). The local Powhatan Indians must have gazed upon these newcomers with not a little chagrin and some mixture of bathos and bewilderment. After all, the Indians knew precisely how to survive; the nurturing of wild game deemed venerable but necessarily consumed. Those exigencies were never pleasant – horrible, if you are a vegetarian – but they were killings marked by some measure of restraint. Indigenous peoples also came to the point of planting beans and broadleaved squash, inviting none of the plant pathogens, beetles, and worms, whose populations exploded in the wake of unnecessary soil erosion, or the overkilling of game, or introduction of *mobile meat* in the form of cattle and pigs and with them contagions like anthrax. These were practices symptomatic of the opportunistic European presence.[5]

---

[4] Kiheung K (2007) The social construction of disease: from Scrapie to Prion. Routledge, New York/Oxon. http://nativeamericannetroots.net/diary/688. Accessed 22 May 2015.

[5] http://www.encyclopediavirginia.org/jamestown_settlement_early. Accessed 21 May 2015

Most striking, the multiplication of Jamestown-like urban plans by the colonists would result in a rectilinear worldview; the forcing of a sacrosanct circle into an ill-suited square of tyranny, rendering mineral, wood, and animal extractions more easily regulated. With arbitrary strictures comes bullying authorities, and with such superiority-cleansing follows desperate class dissent, tribunals, a history of laws and legislation coinciding with equally zealous punishments, warfare, and the distancing if not outright disassociation of people from one another. The end-loser in such antics has always been nature. But the *individual* has also lost much ground, against the backlash of her/his own species because of the inevitable versions of militaristic behavior exhibited by nearly every human society. Seen under a microscope, one would be hard-pressed to distinguish us from other *social systems* in biology, whether ant, termite, red tide, or the dynamics of deadwood.[6]

Indian ecosystem maintenance, however, suggests patterns which today's historians and ecologists continue to debate. Did they know something we modern practitioners of usurpation have forgotten? That uncertainty, with its many theories, likely persists because of contemporary mindsets buttressed by biases of habit, age-old routines that are typical of a perpetual trespass taken for granted, and one that is ill-boding with respect to our own future. In such scenarios is writ large the capacity of individuals to join with like-minded robbers and squanderers, forming a pattern said, in retrospect, to have *tamed the West*: sheer nonsense.

Indeed, as of late October 2016, the tribal chairman of the Standing Rock Sioux had descried the aggression against "prayerful people" as his tribal constituents, as thousands of other protestors from throughout the United States had come together to contest a developer's oil pipeline across territory deemed to be most holy, Native American land. Protestors were being kicked off alleged oil pipeline territories, newly privatized by corporate figureheads and stockholders.

At issue: power. The oldest of clichés, *might makes right*. Human nature has vested authority in power from the earliest documentation of our species. If power is corrupted, human nature witnesses not a collapse but a crisis of conscience. Does such a crisis bode well for the implementation of ethics, let alone a *global ethic*? And is a global ethic even synonymous with a cumulative solution to humanity's devastating inflictions on biodiversity, the underpinnings of any future genes? We have already conceivably lost nearly half of all known vertebrates and invertebrates. Does a human ethic at its very zenith – were it to materialize as our species heads over the brink – invoke rights and duties on our part aimed at invertebrates, particularly in the oceans, the majority of species with the highest numbers? That very phrase, the *majority of species*, hinges on the notion of predominance, of a majority, of power. Can power ever be ethically reconciled by humanity which is itself so torn between classes, religions, and the bewildering hit-and-miss dollar valuations that coincide with nearly everything we do, classify, and think about?

Can humanity reconcile itself with that majority of species? Nothing like that has ever happened, nor do we know if evolution would permit our one so-called powerful

---

[6] http://nationalhumanitiescenter.org/tserve/nattrans/ntuseland/essays/threeworlds.htm. Accessed 24 May 2015.

species to envelop others with a kind of egalitarian-oriented grace. Do human beings have the capacity to outmaneuver evolution, for the benefit of ethics? If so, whose ethics, that of the majority? Or are there inevitable motives, the selfish gene predilections, that would undercut any ethical commons, as was vividly described by Garret Hardin in his December 13, 1968, essay in the journal *Science*, entitled "The Tragedy of the Commons." These questions, each one of them, are stifling.

Take just one of the above queries: Can we alter evolution to suit our ends by putting into practice some version of an ethical commons? Most scientists will be skeptical of this level of altruism – we don't even understand it; we can't envision the paradise sketched by the Prophet Isaiah, or the many beautiful kingdoms throughout literature and art, because the weight of our species' self-interest is so vested that we have never considered the consequences, advantages, and disadvantages of ceding power to Others. In that sense, evolution is classified essentially as God would be classified: greater than ourselves. If we insist on keeping God out of it, then there is only a blind and grasping force for survival that is evolution and to which all genetic information is oriented. That, too, presupposes our belief in the supremacy of genes (much like genetics [read: nature] knows best). It tests the boundaries of human knowledge, or, for that matter, the knowledge of any organism. And this cumbersome train of logic devolves to the notion that knowledge is not equipped to outweigh the forces of evolution (1) and (2) no individual has the power to bypass fundamental laws of nature. None of this makes sense, however, in a world that is not governed by laws, or evolution or cognition or genes, but by something else – much like an *ethical gravity*. If an artist asserts that *something else*, science cannot enter into dialogue with it, except by the most elementary, transparently awkward, and disingenuous of tactics.

When Strangers Meet

Precontact Americans as of some 9700 years ago, the end of the Late or Upper Pleistocene, inherited and/or were to varying degrees naively complicit in the extinction of approximately 75% of all genera in excess of 40 kg, a known 183 species. Fifteen of those genera disappeared in a tumultuous period of a mere 1500 years, at the very end of the fourth glaciation. This Quaternary Megafaunal Extinction ("QME") decimated large mammalian herbivores as well as carnivores: the saber-toothed cat, American mastodon and mammoth, at least three species of equine, a California tapir, a Florida cave bear, a giant polar bear and giant beaver, a Western camel, wolves, lions, cheetahs, and a shrub-ox, to name just some of the most exquisite of beings and all of their kind. Remember, extinction. Every member of a species down to the very last one. Extinction is more than a blitzkrieg. It is the systematic rounding up of every last member of a family, every cousin, distant cousin, even every speaker of their language, or individual who might remember even a word or two, and murdering them. That approaches, in broad strokes, the echoes of extinction multiplied by millions of Jewish Holocausts by the Nazis.

This "prehistoric overkill hypothesis" (one theory among many) puts the burden of this orgy of killing on the Paleo-Indian Clovis peoples. Comparable fatal intersections occurred across six of the seven continents, and on countless islands, particularly in places like New Zealand (newly described as the eighth continent, Zealandia), the Mascarenes, and Madagascar, commencing at least 50,000 years ago. A legacy of destruction across the seventh continent, Antarctica, is being written now, as climate change and the continued mass killing of krill, who are so critical to the food chain, escalate unmercifully by the day.

Few who study archaeology and the vulnerabilities of human nature compare Clovis, Galaz, or Folsom, New Mexico; Plano (Plainview), Texas; or the Mississippian culture Monks Mound at Cahokia, Illinois, to Moscow, Phoenix, Beijing, New Delhi, Sao Paulo, or Jakarta – but they should. The environmental consequences of human nature are too rashly compared with the behavior of our ancestors. Yet, more fur-bearing animals are slaughtered in supposedly environmentally correct Denmark without so much as a veiled doubt, a blink, in 1 year than were killed for (what was then, a part of) human survival across all of North American over half a century during times past by indigenous peoples.[7] This is but one important comparison among countless others.

There is little doubt that the Indians of North America modified the landscape and altered forest composition, plant succession, flood plains, and sedimentation in streams and rivers. By hunting to extinction the largest herbivores during the last millennium of the Pleistocene (in the cold snap known as the Younger Dryas), most emphatically the wooly mammoths, changes in the herbaceous mosaic were unleashed. Short grasslands laced with sweet willows gave rise to forests of birch and marshes of peat, for example.[8] New Mexico's drought-resistant state tree, the desert willow, is actually more like the catalpa. Only its leaves *resemble* a willow.

By the sixteenth century, Native Americans struggled to survive in modes not dissimilar from the Europeans with whom they engaged in various levels of trade and then ruthless conflict. They were farmers, fishermen with expertise in drying meat, others 2000 miles across the continent who could grind acorns into the finest flour and bake spectacular breads, some sweet and razor thin. Others grew tobacco and maize. Wrote Carl O. Sauer "The Indians…roasted and baked tubers in ashes, parched and popped seeds in hot sand, and pit-steamed large quantities of turgid mescal (agave) flower-stalk buds. The basic techniques of cooking are immemorially older than pots, kettles, or even water-holding baskets."[9] These same indigenous peoples knew how to maintain cisterns and had sophisticated irrigation to rival the greatest Medieval Buddhist horticulturalists in Bali. At sites like Pecos, New Mexico, or in large hunter/gatherer aggregates like that of the Yokuts ("people," as translated into English) on what was once an enormous Lake Tulare, 20,000 inhabitants traded

---

[7] http://www.infoplease.com/ipa/A0002130.html#ixzz30sGngTKG. Accessed 21 May 2015.

[8] Native Americans modified American landscape years prior to arrival of Europeans (2011) http://www.sciencedaily.com/releases/2011/03/110321134617.htm. Accessed 21 May 2015.

[9] Seeds, spades, hearths and herds: the domestication of animals and foodstuffs (1952) The MIT Press, Cambridge, MA, p.11.

with other tribes over the High Sierra and lived sustainably for centuries.[10] It took modernism only a few decades to destroy that lake and banish most memories of the Yokuts to the distant margins, from once what were some 60 tribes all speaking the same language to a vastly diminished 7 clans and 8 tribes today.

Throughout the Great Plains, no matter how many bison the Indians hunted, those numbers never came close to an ecological threshold, staying well within the limits any wildlife or park manager of the twenty-first century would deem conservatively sustainable.

Of course, Indian history in North America is not a full story without the paradox of tumult that swept over them. Some tribes fought one another. No person, tribe, or people is without contradiction. We remain hard-pressed to confer upon any one time or place the notion of Utopia, as much as any ecosystem might seem, and perhaps is, perfect. Wordsworth pointed to the inherent contradictions in such thinking, as did Ovid, early Hudson River School, and later Luminist painters, from Thomas Cole to George Inness. However, if we were to view the world through the lens of Jan Brueghel the Elder in the early 1600s (and that of his enthusiastic patron in Milan, Cardinal Federico Borromeo, whose celebration of all that Breughel's paradise scenes added to the Catholic counter-Reformation, published posthumously in 1632 under the title *I tre libri delle laudi divine*), we should come away convinced of that very Eden.

We have always as a species followed upon such perfection with imaginative expression, the prayers and the rituals of sentience. As with the previously described peoples of northern Mexico, we aspire to divine meaning in the human ecological footprint, but with so much variance as to all but negate the notion of certainty as to any predictable behavior. Moreover, we too easily chalk up the varieties of religious and ethnographic experience to demographic anthropology. Writes Tim Flannery, "If a balance ever did exist between Indian hunters and their prey it must have been a fragile one, for changing technology surely challenged it. The introduction of the bow and arrow greatly increased the efficiency of hunting, and circumstantial evidence suggests that in some areas animal numbers were so reduced that Indians began to rely more on agriculture." And Flannery goes on to discuss the cultural and ecological implications of the horse, the gun, and the disease-borne catastrophes undermining indigenous populations in first contact with Europeans.[11]

First contacts are unambiguous, innocent, idealistic, and vulnerable, most famously depicted by William Hodges (who accompanied, as painter, the doomed Captain Cook). But consider what happens on any given day that a person of any species encounters another. It is as if our lives are notoriously or fantastically elsewhere. It is typically miraculous: the first look at a woodpecker, a red panda, a tiger, and a flying fish.

---

[10] Edward DC (1998) Short overview of California Indian History. Cahuilla-Luiseno. nahc.ca.gov/califindian.html. Accessed 23 May 2015

[11] Tim F (2001) The eternal frontier: an ecological history of north america and its peoples. Grove Press, New York, pp 340–341

But first contact has never been without trauma. "A missing teen" in Manchester, United Kingdom, for example. First contact by emergency personnel might mean the tragic confrontation with an unidentifiable scene of body parts and human ruins. As against the corporate, political, legal, philosophical, and animal rights contexts, such personhood categories evince deeply disturbing rifts in the universal and historic story line of what it means to be an individual.

Often, indigenous views of the world are at once accelerated into eschatologies akin to Apocalypse, as was the literal case with the Bimin-Kuskusmin of Papua New Guinea, whose entire world collapsed following the trespass by a few petroleum engineers searching for oil, that turned out to be of great spiritual significance to the Bimin-Kuskusmin. An analogous fate awaited the Ik of Mount Morungole overlooking the Kidepo Valley of Northeastern Uganda, when they were forcibly relocated from their mountain world into an alien flatland of agricultural unknowns.

In the late 1970s, I (Michael) was partially privy to a fiasco of intercession when four Indian military personnel, executing a new survey of portions of the contested McMahon Line between northeastern Arunachal Pradesh and southern Tibet, encountered a dozen, allegedly naked tribal people unknown to the outside world. The individuals were witnessed consuming their food in a cave at approximately 15,000 feet. When these indigenous peoples saw the interlopers, they fled, never to be found again. I (MT) was asked by then Prime Minister Indira Gandhi, as well as by her son, Rajiv Gandhi (soon to become prime minister), to attempt to make contact with the tribe. To parachute (naked) into the quadrant where they were seen and attempt to meet them. The head of a major medical school in India, in collaboration with an American university, hoped that MT would get some sort of genetic sample: a strand of hair or (most unlikely) a blood sample.

It is a complex story that unraveled in February 1983, during the same tragic days of the Assam Massacre. The end result: the tribe remains unknown and has never been seen again. The real story line was this tribe's alleged lack of any knowledge of fire.

Such ill-effects of contact, or attempted contact, continue in those regions of the world today where uncontacted tribes are still believed to exist, particularly in Amazonia. In total, an estimated 100 uncontacted tribes are allegedly out there.[12]

Not dissimilar in dual bewilderment – human and other species – are the very rare encounters over the past half century between humans and newly ascertained species, living or dead, like the recently discovered new wasp group from the Cretaceous found in amber in Burma.[13]

---

[12] Bob H (2013) How many uncontacted tribes are left in the world? https://www.newscientist.com/article/dn24090-how-many-uncontacted-tribes-are-left-in-the-world/. Accessed 31 Oct 2016

[13] Lichao G, Chungkun S, Longfeng L, Dong R (2016) New pelecinid wasps (Hymenoptera: Pelecinidae) from Upper Cretaceous Myanmar amber. Cretaceous Research Volume 67: pages 84–90. Available online 12 July 2016. http://www.sciencedirect.com/science/article/pii/S0195667116301409. Accessed 31 Oct 2016

Or both the smaller Iriomote and larger Yamapikarya wild cats of Iriomote Island, first deciphered in the 1960s, gorgeous and elusive spotted leopard-like beings that have all but vanished into myth (like the newly discovered Borneo leopard). But they are real, indeed, and protected by the Japanese park service, just as Yeti are safeguarded within Bhutan's Sakteng Wildlife Sanctuary. Iriomote is a 111.7 square mile subtropical mangrove forest within the Yaeyama Island group of Okinawa Prefecture, southwestern Japan, 124 miles due west of Taiwan, and replete with the astonishment of these by now legendary encounters. Encounters, one by one, all real. These beings are alive with the verisimilitude of something extraordinary, nuggets of gold in the imagination. As real as that first sighting of a grizzly bear in the North Cascades in October 2010[14] or the discovery in the 1960s of the Tasaday tribe in the southern Philippines, championed by the late John Nance, a tribe whose living legacy has been poisoned by true fake news, though perpetually teased by attribution social science.

Throughout 2015 and 2016, photos and data poured in of rarely seen or newly discovered species.[15] In every such revelation – the first gray wolf pack in California in nearly a century; a pink hippo, a pinkish English grasshopper, or a baby elephant; an albino humpback whale and a similarly rare albino giraffe; a new branch of *Homos* from paleontological digs in Ethiopia; and an unexpected, never-before-seen species in a work of art on a remote rock wall – in each of these events of aesthetic biology, there is a hope we cannot but harbor, akin to entering a new realm of awe and sensibility whence, alas, all is not lost and the human journey is reaffirmed at its most sensitive nerve ending: the connection to Others which in turn fosters ever-increasing evidence of our own true individualism by dint of our ability to celebrate and revere nature and all of her kin.

But in the case of American Indians, or the Armenian people, or the Jews of Europe prior to and during World War II, so much hope in so many lives was indeed expunged. Memory is torn between tenses, mourning the unimaginable horrors meted out, coping with such neural complexities in the everyday present, extrapolating into the future, as the sixth extinction spasm currently sweeping the planet only reaffirms the stakes, underscoring how fragile and miraculous life's myriad biographies and their various biological incarnations and expressions truly are. We exist philosophically trapped by these truths, ensnared while guardedly granted a second chance to reinvent culture, civilization, and our entire biological outlook. An Alberto Giacometti (1901–1966) sculpted one version of this situation, while a Jean-Paul Sartre (1905–1960) described the crisis in a broader sphere by invoking "in itself," "for itself," and "for others" as ontological parameters of fundamental importance to defining human beings, in his masterwork, *Being and Nothingness*.[16]

---

[14] http://www.conservationw.org/news/pressroom/press-releases/first-verified-grizzly-bear-sighting-in-the-cascades-in-fifteen-years. Accessed 31 Oct 2016

[15] https://weather.com/science/nature/news/top-10-new-species-photos; see also http://www.livescience.com/topics/newfound-species. Accessed 1 Nov 2016

[16] L'Être et le néant (1943) Librairie Gallimard, Paris

North American Indian Demographics

When we envision a world populated by Giacometti's various solitaires and walking figures, we are immersed in isolation and, at least by one interpretation, rough and lonely separations. Fated by an inflexible anatomy that is withheld and bereft of ultimate hope, sensing and knowing what is all too real about our mortality, and all too with us about the human world.

That is precisely the ghostly halo that descended like an enchantment transformed eventually into a vast storm cloud over North America's extraordinary concatenation of Indian tribes. In contrasting the sluggish pace of legal protections with the relentless killing and fragmentation of Native American collective life, it is simply a tragedy of astonishing expanse whose ripple effects challenge all future extrapolations in terms of the ecology of human behavior. Psychologists try to cope with such questions as whether millennials, for example, are more empathetic toward animals than their children or their parents; whether anything other than the sky falling will actually shift human cultural mores. And even with a falling sky: what constitutes that level of gravity?

With respect to American political maltreatment of indigenous cultures both at home and abroad, despite the peace initiatives approved by Congress starting in 1867 (the Treaties of Fort Sumner, of Medicine Lodge, etc.), the Indians were not given anything like the legal loft intimated in legislation for all other residents of the United States. Write David Hacker and Michael Haines, "Over the course of the twentieth century, researchers have estimated the Indian population of the coterminous United States as low as 720,000 (Kroeber, 1939) and as high (for all of North America) as 18 million (Dobyns, 1983). Most estimates fall in the range of 2–7 million, implying a population loss between 1492 and 1900 in excess of 85 percent."[17]

The US Census for 1900 revealed that Indian populations had dwindled to 237,196 individuals.[18] They were confronted by more than 75 million non-Native Americans, spanning all 46 states, at that time. This demographic nightmare mirrors all the current patterns of the Anthropocene, placing human life in the same near abyss as most other megafauna on the planet, from sequoias to blue whales, from Amur tigers to polar bears.

By the time of the Great Depression, the Indian's near invisibility – by percentage – was rendered even more unstable and jaundiced in the eye of most Americans by the fact that the Indians were, by and large, legally landless (laws other than their own, imposed with a retaliatory determination), with no means of even the most

---

[17] David JH, Michael RH (2005) American Indian Mortality in the late nineteenth century: the impact of Federal Assimilation Policies on a vulnerable population. Annales de Demographie Historique n 2:p. 17 à 45 p.19

[18] Leonard CS, James GR (2003) Historical Dictionary of the Gilded Age, ME. Sharke, Inc., New York, p.332; see also Russell Thornton (1987) American indian holocaust and survival: a population history since 1492, University of Oklahoma Press

marginal subsistence farming, a crisis of cumulative oppressions scarcely lifted prior to the enactment of the Alaska Native Claims Settlement Act (43 U.S.C. 1601 et sec.) in 1971.

By then, Indian assimilation into American culture was measured according to those from whom taxes could be extracted, who wore "white-man's" clothing and studied English in schools. As early as 1865, a special Congressional committee determined that the attrition of Indians could only be reversed by civilizing them. While Congress began focusing upon vaccinations for American Indians in 1832, the persistent neglect of all those tribes west of the Mississippi was not even remotely rectified until 1955.[19]

In 1885, Congress had deemed land to be a key factor in resurrecting Indian self-maintenance. The Dawes Act allowed for allotments of 160 acres to each Indian family head. Individuals over 18 received 80 acres, while minors received 40 acres, held for one generation in trust by the federal government. By 1900, 343,351 acres had been allotted, which translated into roughly 1.3 acres per Indian left in North America,[20] spread across approximately 2500 Bands, Gens, and Clans, as analyzed by Frederick Webb Hodge.[21] This was the same Hodge who would edit Curtis' masterpiece. These were the same data, the vastly attenuated tribal lands, that made Curtis' effort to document those who were left, on their land – after so combustible an ecological history – the most important, and at the time seemingly implausible, human rights initiative of the twentieth century.

By 2013, the US Government recognized 566 Indian tribes. Curtis had documented some 80 of those.

To put all of this in context, prior to Columbus, there were as many as 1000 Indian languages, 250 in what would become the United States, and as many as 112 million indigenous peoples, though most estimates are far more conservative, suggesting approximately 8 million. Regardless of the pre-Columbian numbers – we will never know, for sure, an astonishing gap of knowledge betrayed by a vast cadre of academic witting or unwitting conspirators – what is far less ambiguous is the fact that by 1650, there were fewer than six million Indians remaining in North America, as a cascade of intensively documented massacres quite literally undermined tens of thousands of years of post-and-pre-Clovis cultural evolution.[22] As William M. Denevan – the Carl O. Sauer Professor of Geography and Environmental Studies at the University of Wisconsin–Madison – has written, "The discovery of America was followed by possibly the greatest demographic disaster in the history of the world."[23]

---

[19] See Christine Massing's (1994) The development of United States Government policy toward indian health care, 1850–1900. Past Imperfect. 3:129–158

[20] Otis DS (1973) The Dawes Act and the allotment of Indian lands. University of Oklahoma Press, Norman [1934], p.139

[21] 1905 Handbook of American Indians North of Mexico: A-M, Frederick Webb Hodge, US Government Printing Office, 1907

[22] http://uwpress.wisc.edu/books/0289.htm. Accessed 23 May 2015.

[23] William MD (1992) The native population of the Americas in 1492, 2nd Revised Edition. University of Wisconsin Press, Madison, Wisconsin. http://uwpress.wisc.edu/books/0289.htm

Enter Edward Curtis

Curtis, the son of a farmer/minister, was born in Whitewater, Wisconsin, on February 16, 1868, less than 6 months before the passage of one of the most hotly contested pieces of American legislation, the 14th Amendment to the US Constitution, ratified on July 9, 1868. By that action the abolition of slavery (earlier enshrined in the 13th Amendment of 1865 and further enhanced by the 1866 Civil Rights Act) gained significant traction; fundamental details of everyday life were cemented into law, and the heinous Supreme Court Dred Scott decision of 1857 was forever banished. The 14th Amendment, by its Equal Protection Clause, and declaration that States could no longer "deprive any person of life, liberty, or property, without due process of law" suggested to the idealist – and most likely to Curtis' evangelical father – that America had finally enacted a universal conscience, a "peaceable Kingdom."

Environmental ethicists and philosophers have pointed out that this 14th Amendment was a central pillar of American rights and freedoms, a clear forerunner of the modern environmental justice movement, most notably enshrined nearly a century afterward in the Civil Rights Act of 1964,[24] the same year as legislation confirming America's first Wilderness Act and National Wilderness Preservation System, "where man himself is a visitor who does not remain." A visitor. A visitor from where? A visitor to what? Does it connote an entire world humans may but visit? And by that logic, what are humans? Who are these visitors? A species that found it in the best interest of the world – or at least those fragmented pristine remaining portions of the world – to make itself a visitor.

Civil rights and environmental rights, animal rights, and plant rights. How do these stack up against taxonomic history, simplistic binomial nomenclature given to the fundaments of genus and species and the various legislations, from nation to nation, in addition to the nearly 900 acts conferred upon the rights, duties, and circumspections of human behavior by international ecologically related treaties?

Various provisions of the 14th Amendment, particularly those contained in Title VI 601 and 602, would form the legal basis for numerous successful actions in the 1970s and 1980s involving what the Supreme Court would come to view as racism in the guise of environmental injustice, deliberately unfair treatment by states or municipalities of minorities with regard to such rights as access to city parks or drinking water or sanitation services.

But a century prior to those environmental rallying cries, at the time Curtis was born, there was no Universal Declaration of Human Rights; no Earth Day or EPA or Charter for Compassion; no UNEP, UNESCO, UNICEF, or the International Court of Justice ("ICJ," also known as the World Court) in the Hague; and no line of defense to prevent or even address genocide, let alone ecocide. Not a single international environmental treaty with any binding powers.[25]

---

[24] American Environmental Justice Movement, Internet Encyclopedia of Philosophy, http://www.iep.utm.edu/enviro-j/; Eddy F. Carder, Prairie View A & M University

[25] Data Sources for International Environmental Politics. https://rmitchel.uoregon.edu/data1, University of Oregon. Accessed 1 Nov 2016

As the young Edward Curtis, whose family had moved to a small town in Minnesota where Curtis dropped out of grade school and built his own camera for $1.50, was soon to discover, the 14th Amendment did nothing to protect the American Indians. By the late 1860s, despite varying degrees of concern demonstrated on behalf of Native Americans, the Indian's collective world was verging on extinction. The reality was palpable, not only to Indians.

Outspoken individuals had certainly, from time to time, protested or portrayed the ill-treatment of indigenous peoples, but all such pleas or artistic nostalgia were insufficient to impede the ongoing tragedy of manifest destiny, particularly with the advent of railroads. Genocide deniers of the nineteenth century were equal in so many patterns, economic motives of self-interest and patois to climate deniers within the petrochemical industries and politics of the twentieth and twenty-first centuries.

Ralph Waldo Emerson had written a letter to President Martin Van Buren in 1838 in regard to the forced relocation of 16,000 Cherokees east of the Mississippi River under terms of the Indian Removal Act of 1830, bullied into law by Andrew Jackson. That river had become a veritable demarcation zone between what most Americans and politicians in Washington thought of as the red line separating the wild (savage) West from the civilized East (and, nearly a century later, it would represent yet another form of demarcation between Whites and suffering Blacks in the 1927 musical, "Ol' Man River"). Wrote Emerson, it is "a crime that really deprives us as well as the Cherokees of a country; for how could we call the conspiracy that should crush these poor Indians our Government, or the land that was cursed by their parting and dying imprecations our country, anymore?"[26]

Henry David Thoreau went to jail (for a night) rather than pay taxes to a government of oppression against which he would ultimately – in the case of Captain John Brown – show support for not mere civil disobedience but outright violent resistance, thus, in his own manner subscribing to the desperate struggles to resist American sentiment that largely chose to ignore the Trail(s) of Tears or the final armed inflictions by US military upon Indians.

Other earlier opponents of the US conquistador mentality included the great ethnographer Frank Cushing (1857–1900), who would become the first "outsider" to live with, and be honored by, a Native American tribe, the Zunis of New Mexico. Cushing had lived among them between 1879 and 1884 and been affectionately introduced to their most intimate spiritual understandings. He was given a Zuni name of Tenatsali, "medicine flower," wrote at least 14 books, and introduced a style of ethnography that was participatory and experiential, not remote and objectifying. Cushing penned works extensively on the Hopi, as well, and led the Smithsonian's major archaeological discoveries in Florida of ancient Indian dwellings (major sites today covered by buildings and parking lots). But when a land grab by politically well-connected Easterners encroached on Zuni territory, not even America's most famous anthropologist, Cushing, could do more than write outraged

---

[26] Joel M, Len G (1995) Emerson's Antislavery Writings. Yale University Press, New Haven, p.3

editorials that failed to help his Indian friends' cause, no matter how seemingly persuasive his ardor.[27]

Similarly, lawyer, painter, and author George Catlin (1796–1872) would become enamored of Indians and their spiritual legacy, exploring much of the Plains Indians' worlds, painting over 500 exquisite oil portraits of their nobility and poignancy (as painter Charles Goldie had done with the Maori of New Zealand). Catlin was a precursor, to be sure, of Edward Curtis.[28] From 1831 to 1837, Catlin traveled with the Pawnee, the Creeks, the Lakota, the Comanches, and other tribes, endeavoring to make accessible their exquisite personages, their ecosystems, and their varied manner of existence to an American public. Peter Matthiessen would characterize Catlin's heroic endeavor – paintings, journals, and his two-volume *North American Indians* executed with a verve akin to other great and interested painters of the time, like the romantically inclined Alfred Jacob Miller and, later, Henry Farny – as the first and last true record of the Plains Indians recorded at the zenith of their regal and sustainable lifestyles.

Catlin actually called for a "national park" to be created by Congress to protect Indians. But this appeal was promulgated just prior to those Indians' destruction, by means that the late Matthiessen described as "traders' liquor and disease, rapine, and bayonets."[29] Catlin's concept invoked what modern conservationists have called "inhabited wilderness," the notion that "rewilding" with lions, elephants, giraffe, camels, and the like would not only restore ecosystems but had a chance of restoring the vitality of Native American livelihoods in accord with their spiritual and sustainable past.

In Catlin's 1868 book *Last Rambles Amongst the Indians of the Rocky Mountains and the Andes*, he looked back over more than a quarter-century and wrote, "…The eighth day opened to our view one of the most verdant and beautiful valleys in the world; and on the tenth a distant smoke was observed, and under it the skin tents, which I at once recognized as a Crow village." He had much later depicted that scene in his 1855 painting "A Crow Village and the Salmon River Mountains" (Mellon Collection, National Gallery of Art, Washington, DC).[30] This is reminiscent of one of the first members of the US Cavalry who, upon seeing Yosemite Valley from a distance, is said to have broken down and wept.

In a similar vein, Albert Bierstadt's "The Rocky Mountains," 1863 (Metropolitan Museum of Art), and his "Sunset Light, Wind River Range of the Rocky Mountains,"1861 (Free Public Library, New Bedford, Mass.), yielded memorable windows on the great spiritual and ecologically instinctive life of American Indians, views that have become subsumed within the penitent aesthetic that today infuses so much of the unspoken realization that a genocide occurred in our very midst, a tragedy both human and involving all of nature.

---

[27] http://www.pbs.org/weta/thewest/people/a_c/cushing.htm. Accessed 24 May 2015

[28] http://americanart.si.edu/education/pdfcatlin.pdf. Accessed 24 May 2015

[29] George Catlin (2004) North American Indians, Introduction by Peter Matthiessen. Penguin Classics, New York

[30] In Patricia Trenton and Peter H. Hassrick (1983) The rocky mountains – A vision for Artists in the nineteenth century, University Of Oklahoma Press, Norman OK, p 281

That similitude of ecosystems and the people most directly dependent on the biotic heritage of North America surfaced in other depictions: the violence suggested in Charles M. Russell's "Lewis and Clark Meeting the Flatheads at Ross' Hole, 1912 (Montana Historical Society, Helena, Montana), and Frederic Remington's "Fight for the Stolen Herd," ca. 1902 (Private Collection).

In sum, the Arcadian brush of Hudson River School painters or sincere appeals of Transcendentalist poets fell on the ears of a deaf century as America accelerated the annihilation and assimilation of American Indians. By December 29, 1890, as more and more Plains Indians had come to embrace the mystical call of the Paiute man, Wovoka, who had spread the religion of the Ghost Dance, and articulated what he perceived as the End of the World, that dark vision would engulf innocent victims of US military madness on the icy fields at Wounded Knee Creek, South Dakota.[31]

It has been said that Wounded Knee evinced the death knell of the American Indians. Their ecological wisdom, it would seem, had backfired in that their very success had invoked a complex psychoanalytic minefield of competitive disadvantage against so myriad and multitudinous an opponent, aped on by the cumulative realities of the Louisiana Purchase, the Lewis and Clark Expeditions, and the like. But in the tragedy of the slaughtered Lakota Sioux lay the foundations for a great artist to ennoble Indian history with a recollection steeped in admiration, love, and fidelity to an ecological and spiritual way of life that holds the promise of tutoring our own future.

All that embittered, racist opprobrium of past centuries would be edified by a new resurgence of admiration, even nostalgia, as galvanized in the work of Edward Curtis. His aesthetic, courageous advent bolstered some other set of forces in opposition to the old fears and enmities. He captured the nobility of Native Americans but also their powerful, inspired innocence. He stared directly into the eyes of our last hope and opined on the possibilities for our future. In Curtis was the embodiment of an ethnographic theory: that all things hypothetical are possible.

This latter trait, this hope that runs through all of Native American history – an innocence in the best sense of the word – holds an important clue to the clash of cultures and the reconciliation that a great artist like Curtis can engender.

In the recent restoration of the 1494 "Resurrection" fresco by Bernardino di Betto's (known as Pinturicchio) hanging in the Vatican's Hall of Mysteries, a clear focus on a previously hidden detail has emerged, one that suggests possibly the earliest depiction of Europeans with Indians. Think of all those 2015 photographs, videos, and discoveries of new species. Seven Indians, naked, are – by all appearances nobly serene, Apollonian, peering inquisitively half – hidden behind one another. One or two stand proudly out front, with incisive intelligence and gentle demeanor, bearing parrot head-feathers, one man holding a staff, right out of Tamaulipas, northern Mexico, circa BC 5000. The painting is instinct with awe and sheer wonderment, particularly on the part of one soldier, enlarged, in the foreground, who gawks upward toward the Christ figure in the painting, but not without grasping the ballet of delight directly below, this congeries of enigmatic and beautiful

---

[31] http://www.lastoftheindependents.com/wounded.htm. Accessed 25 May 2016

New World ambassadors.[32] A similar sensibility is conveyed by the Dutchmen, such as Henry Heusken who negotiated for several years on behalf of Americans new to Japan and helped enable the first commercial treaties between the United States and Japan in the 1850s. His diaries, spanning the period 1855–1861 and published more than a century after his murder (in Japan), make clear that he considered the Japanese at the imperial court to be the most civilized, sophisticated, and elegant of any people he had ever encountered or even imagined. His *diaries* are a remarkable, deeply moving reading experience.

Centuries after Pinturicchio's time, such encounters might well morph into a violent trespass, as witnessed in the many renditions done of Captain Cook meeting for the first time indigenous Hawaiians. Or, in today's artistic translation, such memorable and horrid encounters as depicted in films like director Roland Joffé's "The Mission" or Mel Gibson's "Apocalypto." In 1494, however, the anthropic conflicts had not yet erupted. An original astonishment, an aboriginal nature was akin to religious enlightenment felt by one and all.

The Revelation in Seattle

By 1885, when the 17-year-old Curtis and his struggling family set out for Seattle in hopes of securing paying jobs, the annihilation of the Indians was continuing full-force. Curtis obtained a new camera and got involved in one of the first photographic businesses in the Northwest. He was, on a multiple of levels, the US Justice Department, CNN, the Internet, FedEx, and the telephone, all wrapped up into one artistic conscience. He stepped forward to deliver an urgent message to every American.

A remarkable image of Seattle around that time, "Bird's Eye-View of the City of Seattle, Washington," in 1884, drawn by Henry Wellge, speaks volumes of the rapidity of change overtaking the United States and the city into which Edward Curtis would gain notoriety and inspiration.

Seattle's population had soared from 1107 in 1870 to 42,837 in 1890. But by 1884, the image of the city resembled a massive array of black-smoke-belching monster ships in Puget Sound, due to it having, by that time, at least 15 major manufacturing establishments, including metal casting and machine shops, saw mills, gasworks, galvanized iron cornice works, and furniture and barrel makers, 3 major banks, and at least a half-dozen hotels. It was already a big city. There was even an opera house and, as a historian of 1890 reported, "Seattle was in the midst of its transition from a frontier town to a great commercial city." Amid this daily riotous transition, "three murderers were lynched in the public square."[33]

---

[32] Hrag V (2013) First Western Painting of Native Americans Discovered at the Vatican. http://hyperallergic.com/70448/first-western-painting-of-native-americans-discovered-at-the-vatican/

[33] John WR (1979) Cities of the American West: a history of frontier urban planning, Princeton University Press, p 384

Towering beyond Seattle was Mount Rainier, of course, first ascended in 1870. Curtis climbed it repeatedly, principally with the Portland-based mountaineering club, the Mazamas, which had named Curtis an honorary member beginning in 1897. John Muir was another honorary member. In the summer of 1897, Curtis led 87 people toward the summit of Rainier. Fifty-nine climbers made it. One, a Professor McClure from Portland, fell to his death. The national magazine, *Harper's Weekly*, made mention of it, naming Curtis as part of the story. The following year Curtis published a portfolio entitled "Scenic Washington," featuring Rainier and other scenes of natural wonder. Such endeavors led to his fateful guiding of George Bird Grinnell up the not infrequently deadly mountain, at which time he is said to have essentially rescued two scientists lost amid Rainier's labyrinth of sweeping icefalls: Clinton Hart Merriam and Gifford Pinchot.[34] With its 26 glaciers, the 14,411 foot Rainier is considered to be one of the most deceptive volcanoes in the world, for climbers, a labyrinth of crevasses and hanging glaciers. But it certainly set well with the burgeoning photographer that was Curtis.

But this mountaineering contagion and love of landscape photography was soon overtaken by a far more pressing emotional realization by Curtis: the fast escalating plight of the last remaining Native Americans.

In the quiet solace of time spent on a beach along Elliott Bay, West Seattle, with the last surviving daughter of Chief Seattle in 1895 – Kickisomlo, or "Princess Angeline (c.1800–1896) – Curtis took numerous images of this elegant and poignant elder, paying her one dollar per picture. He declared that she seemed to prefer earning money that way, than by the back-breaking work of digging for clams. Much can be read into this remark. Nor were these pictures by any means the first portraits of a Native American, as has often and mistakenly been claimed. Earlier photographers like Charles DeForest Fredricks (1823–1894) and the prolific William Henry Jackson (1843–1942) had been photographically enshrining Native Americans for decades.[35] From his studio in Bishop, California, Andrew A. Forbes had focused extensively on the documentation of the Owens Valley and Yosemite-Mono Lake Northern Paiute tribes between 1903 and 1916 with great success and deeply sensitive intelligence.

But Curtis was a very different eye, attitude, and passion than any of his photographic predecessors or contemporaries. His brand of alleged *noble savage romanticism* was rueful, deeply aware of the gravitas enshrouding vanishing cultures to whose many homes he'd been invited. His goal, as expressed in so many images and words, was to reveal nobility much like Henry Heusken had attempted, in describing the Japanese at court. Velázquez' "Las Meninas" and the many portraits from the seventeenth-century Dutch Renaissance by Johannes Vermeer, Gerard ter Borch, and many of their contemporaries come immediately to mind.

---

[34] See Edward SC, the North American Indian, Incorporated, by Mike Gidley, Cambridge University Press, 1998, Chapter Two

[35] Massachusetts Historical Society, Photographing the American Indian: Portraits of Native Americans, 1860–1913, from the collections of the Massachusetts Historical Society". http://www.masshist.org/photographs/nativeamericans/essay.php?entry_id=72, 26 May 2015

Princess Angeline died a year after Curtis photographed her. Two of his images of Angeline digging clams and gathering mussels were chosen to be part of a National Geographic Society photographic exhibition at the time. A third Curtis photograph, "Homeward," depicting Indians heading toward shore on Puget Sound at sunset in a traditional canoe, was also selected and went on to win the Society's Grand Prize. That, combined with the legacy of Angeline and her famous father, catapulted a Curtis trademark combining rural sensibilities, the midwestern preacher/farmer's son, famed alpinist, and photographer-turned-environmentalist. It was a welcome pedigree in a city and a culture being fast enveloped by greed, polluting industries, and the onrush of early modernity.

No one knows what Chief Seattle – Angeline's father – said or didn't say. There were countless versions of his famed ecological coda, at least three layers of translation, and much potentially apocryphal attribution. Nonetheless, the city was named after him and his last surviving daughter's connection to Curtis would prove monumental to the future cultural and spiritual understanding of Native Americans, not just in the United States but worldwide.

This alchemy all came about in a whirlwind, if one looks back upon Curtis' amazing journey. It was legendary, and the city of Seattle mirrored the very history and epic convolutions that were overtaking all of North America. But what stands out amid this flurry surrounding Curtis were the many individuals he sought successfully to portray. Real people, real accoutrements, real contexts. A reality that comports with a new kind of twentieth-century philosophy, one that is writ perplexingly in the myriad mysteries, woes, enchantments, and bewildering contexts engendered three centuries earlier by the likes of Shakespeare's Caliban (from his last play, "The Tempest") and throughout Cervantes.

By July 31, 1899, the world's first underground hydroelectric plant went online just outside Seattle at Snoqualmie Falls delivering power to the nearby metropolis.[36] That's how close to the avalanche of modern technology Curtis' tribal individuals were, a superimposition of mores and clash of cultures that aptly symbolizes the tragedy of progress as it continues to inflict degradation, loss of habitat, and vanishing languages upon the human world. No one quite captured this hybridized future in the here-and-now like Edward Curtis.

The combination of Curtis' thriving photographic business and his new high-powered group of scientific friends led him, first, to Alaska, on the Harriman Expedition for 2 months during 1899, and then to Montana in 1900 at the invitation of his new friend, George Bird Grinnell. These biographical events have been exhaustively studied in many superb biographies. But two salient points emerge from their vortex. First, Curtis had come face-to-face with that long ecological history of American Indians rising to an emotional zenith in the poignant guise of Princess Angeline. And then, in the Algonquian people, the Piegan or Piikáni – the Blackfeet – Curtis was to emotionally and artistically sense the *soul* of an entire tribe.

---

[36] http://www.seattleweekly.com/1999-07-21/news/best-of-seattle-1899/; see also http://seattletimes.com/news/local/seattle_history/articles/timeline.html, 26 May 2015

One of Curtis' photographs from that expedition, "The Three Chiefs, Blackfoot, Montana, 1900," would sell at Christies in 2007 for $115,000, one of the highest prices ever paid for a single Curtis image. And it is thoroughly understandable: In those distant, elegant men on their horses in the short-grass prairie, stopping for water, their thoughts seem to imbue posterity with the long, complex, and lyrical relationship of three philosophers (as Giorgione might have conceived of them) out in the world. It is a quintessentially meditative work and embodies the great spiritual transformation occurring in Curtis himself at that time.[37]

The image resonates with intense and enduring viability. Grinnell and Curtis had had much time to discuss Grinnell's grave concerns about the fate of the Indians, as well as the buffalo and all other species across the Great Plains. Curtis had become a total convert to the realization that Indians were, indeed, vanishing and disappearing with a horrifying alacrity. This was the beginning of Curtis' 20-volume masterpiece, with its large portfolio of photogravures, *The North American Indian*. A known 222 sets would ultimately be published.

Curtis' Great Irony

Curtis could not easily elude certain outright contradictions. Almost everyone who helped him shape his career – everyone other than the Indians – exerted enormous influence over all of American culture and media, had great wealth and/or prestige, owned vast estates and industrial and banking empires, ruled Washington (as in the case of President Teddy Roosevelt), and lived in ways that were all but antithetical to the all-poignant ethos, that very ecological and spiritual heritage his own deeply personal aesthetic was attempting to convey in every portrait of every Indian he ever had the privilege to meet, converse with, and photograph. A by then devoted student of nature, Curtis found himself indirectly indebted to the steel industry, the railroads, presidential politics, and even – fleetingly – Hollywood. His Zen and his Tao were commercialized. Posterity, at first, would ignore and then hold in suspicion his motives, methods, and aesthetic. Within very few years, his body of work would be ignored and nearly lost to history. The artistic and anthropological critiques of Curtis, prior to his rediscovery in the early 1970s, all stemmed from a continuing racism from without, and residual terror from within, that has so marginalized American Indians to this day.

When J. P. Morgan acquired the National Steel Company from Andrew Carnegie for $492 million ($13.95 billion in today's dollars), it was the first billion-dollar

---

[37] Lot 349, http://www.christies.com/lotfinder/photographs/edward-s-curtis-the-three-chiefs-blackfoot-4972170-details.aspx; See Cardozo (1993) Edward S. Curtis: Native Nations, Bulfinch Press Little Brown & Co., p 33; and Cardozo (2000) Sacred Legacy: Edward S. Curtis and the North American Indian, Simon & Schuster, p 30. The image also seems an ironic prelude to the 1940 Oslo publication of Paulus Svendsen's book *Gullalderdrom og Utviklingstro, The Dream of a Golden Age and the Belief in Progress*. See "The Noble Savage Until Shakespeare," by Gösta Langenfelt, pp 222–227 | Published online: 13 Aug 2008, https://doi.org/10.1080/00138385508596950. Accessed 9 Aug 2017

company in history, capitalized at what today would be nearly $40 billion. And when Harriman got back from Alaska, he was to retreat to his 40,000-acre estate in New York, with a home, the "Arden Estate," sitting comfortably across nearly 100,000 square feet of interior.[38] This was a far cry from a tepee. At his death in 1909, Harriman's estimated fortune hovered around $100 million.

As for Grinnell, though the son of a wealthy New Yorker family, he had turned his jejune finances into ecological and ethical altruism. Grinnell would prove to be one of the most persuasive, knowledgeable, and genuine lovers of wildlife – pivoting these passions on a public platform – in American history. And there was no other individual who did more to help the Plains Indians than this humble zoologist, ethnographer, explorer, and author. His friendship with Curtis proved to be the turning point in Curtis' life and career.

When in 1900 Curtis joined Grinnell on that visit to the Blackfeet (aka "Blackfoot") – who numbered approximately 20,000 in that year – it was emotionally transformative for the Seattle photographer. He now came into the realm of a Native American spirituality that viewed all life forms as sacred and interconnected. And he also realized that he himself was a member of that class of human beings that was driving Indians to the brink; seemingly good-hearted Americans who teamed up and down the broad avenues of the new urban reality that was overtaking North America. The contradictions were heartbreaking for Curtis, and we see it throughout his body of work. Curtis' 2 weeks on a Blackfeet reservation were pivotal, enabling Curtis to crystallize his vision for a much more ambitious project that would – via the influence of President Roosevelt – lead him to the doorstep of J. P. Morgan. The rest is history.

The Rembrandt of Photography: Edward S. Curtis and America's Environmental Social Justice Movement

As can be readily ascertained from even the most cursory perusal of Curtis' images (and most gorgeously curated in a concentrated book by Christopher Cardozo,)[39] the young Edward S. Curtis became what we would think of as the Rembrandt van Rijn of Native American portrait photography. Like Rembrandt, whose courage emerged as he aged and was resolved to show what it was like to be a real human being, Curtis photographed the reality of Native American life. If he chose to show that life in a certain light, it was because – we hold – that he understood the multifaceted dignity that attended upon the Indian's unwavering commitment to truth. Amid a welter of confused, besieged, sublime, and often ruinous verities, the American Indians Curtis had the great honor to know presented a window on the possibilities of human nobility and honesty, in a world more readily seduced by power, swayed by greed, and committed to a rabid, time-worn egotism.

---

[38] www.theardenhouse.com/home.html. Accessed 25 May 2016

[39] See Cardozo (2015) Edward S. Curtis: One Hundred Masterworks, Prestel Publishing, Random House, New York

Christopher Cardozo has speculated that probably some "90%" of all of Curtis' 40,000 + negatives "include people either as portraits or 'peopled landscapes'. The remaining few percent are either pure landscapes or still lifes. There are also a number – perhaps some 1% – that are ceremonial images with people in masks or other accoutrements…"[40] This would suggest that of the more than 80 tribes Curtis visited and photographed from 1895 to 1928, between 30% and 50% of all living Native American families west of the Mississippi River would/will recognize someone, some family tie, some distant connection. These percentiles track freakishly with the numbers on the IUCN-Red List, that of the world's threatened organisms, currently at more than 79,000 species that are in danger of extinction.

By his deep artistic and social conscience, Curtis most assuredly helped resurrect an otherwise "vanishing race."

And although Curtis all but gave up his photography after 1928 until his death in Hollywood in 1952, by the 1970s, Curtis' body of work began quickly to be fully "realized" in the American soul, rivaling in value, scarcity, and sanctity the publication of *The King James Bible* and John James Audubon's *Birds of America*. With the 150th anniversary of Curtis' birth upon us in 2018, his magnificent achievement – this lonely, piercing, courageous, and joyful paean to many tens of thousands of years of Indian spirit, heart, and legacy – represents for us personally the greatest ecological and artistic insight into North American history that has ever been articulated. Curtis is, without a doubt, the father of the environmental social justice movement, as it is known worldwide today – by lawmakers, artists, historians, politicians, psychologists, environmentalists, educators, people of faith, and all those members of humanity clamoring for fairness, decency, and human rights.

Looming over his remarkable achievement is that which tempts human destiny: That noble personage squarely intimated from within the vast erratic and self-destructive nature of humanity's collective choices thus far. We take solace within the whole realm of native American venerability such that, finally, one native individual known as the 9000-year-old "Kennewick Man" has been granted the right to be laid to rest in Washington State's Columbia River Basin Plateau, where his bones were discovered, under the Water Resources Development Act of 2016, "which includes the 'Bring the Ancient One Home Act'."[41]

---

[40] Personal Correspondence with Christopher Cardozo, early May, 2014

[41] Kennewick Man Coming Home (2016) http://www.mycolumbiabasin.com/2016/12/12/kennewick-man-coming-home/. Accessed 15 Dec 2016

Chapter 4
A Genetic Cul de Sac

How the Individual Reshapes the Species

The vocative proceeds the nominative; the person is a categorical concept that is theoretically prior to the object. Linguistically, in human terms, at least, the verb exists on account of an individual, of a noun. The noun is at once an action, a qualified subject that (who) is responsible for the action.

That subject need not be alive, however. For a live subject is always capable of being transformed into some version of action, and hence, a verb, stemming from *something* that is dead or alive. But the very concept of *individuals*, from a genetic perspective, is far less open to transformation. Individuals are genetically blurred in the vast seas of cellular mitosis and meiosis, the very production of eggs and sperm. But it is not our intention here to recross the borders long established of genetic frontiers wherein the resulting entity is the by-product of genes. Fertility is not simply a question of male and female, or simple cell division. There are countless varieties documented throughout both the natural and unnatural world that result in the creating of a being. The varieties of the DNA helix are endless. How much we don't know is staggering. The narrative is necessarily tautological because we don't know what we don't know. And what geneticists and molecular biologists *do* know suggests such a dazzling variety of possibilities for the formation of a life form – from potato to bumblebee – that it remains an entirely open question to pose *meaning*, philosophical or even practical meaning, in the guise of a singular personhood. For this reason alone, the biological sciences have been at odds with the *individual* since the very inception of applied generalities concerning species.

As much as we may laud the rare Edward Curtis for his poignant resurrection of "a vanishing race," our current understanding of the Creation is so deeply limited as to excite a thorough reaffirmation of our naivety. With it comes the promise of new behaviorisms commensurate with one's own deeply personal realization of cognitive infancy. That childhood is not unakin to every wild enthusiasm associated with biophilia and its prospects for the coming days and years. Our species' humility

© Michael Charles Tobias and Jane Gray Morrison 2018
M.C. Tobias, J.G. Morrison, *The Theoretical Individual*,
https://doi.org/10.1007/978-3-319-71443-1_4

should be characterized first and foremost as a matter of the appropriate confrontation with the reality we inhabit and of that which surrounds us on all biochemical sides.

That said, the degree to which a human individual is willing to cede her/his supremacy in this matter to the greater good, the surrounding wildness that is the true world, marks the potential beginning of a new nature, as we have called it. A nature that is prepared to dispense with species boundaries and focus instead upon the responsibilities, duties, and potential of the individual. Duties toward one's own family, clan, tribe, species, genus, etc. – to the Earth at large. Put differently, the perennial expansion of our empathy and tolerance is the only antidote to human narcissism. With more than an eye toward, but an entire affability of cognition, unhindered by solipsistic obsessions, unfettered by self, free to carouse as Walt Whitman celebrated in his *Leaves of Grass* (1855). His "Song of Myself" embodied within that work has been condemned by some literary critics as American literature's foremost exercise in egotism. Yet, from an ecological perspective, it is a paean to zoological pantheism whose trajectory across the entire Western Hemisphere commences from the voice of an individual. There are no obligatory launching pads for mythopoetics. They erupt in every biologicalally.

Our blinders in these regards come quickly into focus. For example, those who would debate the viability of a human fetus, while ignoring the slaughter of well over 50 billion vertebrates per year for human consumption, sheds light only upon a huge and lethal hypocrisy. Not all people and cultures are so blinded, of course. The Jains acknowledge life in every grain of sand, dew drop, detritus, and other geological fragments. Taoists worship every rock. Animists of every persuasion see life in everything. For us to focus upon the fundamentalist human origin as a means of persecution of females is sick and sadistic. Jains and Taoists, Brahmanical Hindus and Buddhists, and many others from around the world follow upon the animistic sentience with an equal reaffirmation of the life force in every living being, formed, unformed, child, adolescent, adult, human, cow, turkey, pig, tiger, tree, invertebrate, quadruped, bird, etc. That does not mean they have all relinquished self-defense, or the impulse to come to the aid and lend assistance to others. Quite the contrary.

The Jains, for example, recognize five senses and have taken it as a matter of non-violent compromise to impose upon only those beings with one sense (e.g., various fruits, grains, and vegetables) for their nutritional consumption. Jain ecological strictures circumvent the notion of delectation or passion, preferring instead to recognize the superlative importance of utter restraint in all dealings.

Under such budding circumstances of human conscience, there is no debating the sacrality of life, nor the relevancy of a dignified death. What matters most is the degree to which the individual rises to an occasion and consistently, assiduously strives to help others. The relational insistence is that which leads one through the labyrinth of contradictions and away from the genetic cul de sac that scientific clarity imposes when it constrains our viewpoint to that one species which is said to be doing all the viewing, a ludicrous boast, of course.

Take *Branchiostoma lanceolatum*, the eel-like lancelet, a marine invertebrate found on shorelines throughout the world, as but one modest example of another being whose life histories surely rivals in every relevant respect those of our own kind.[1] Instead of what we think of as a head, they have a mouth, adjoining their gills and thereby efficaciously amplifying their ability to easily capture their meals. At the same time, not unlike humans, they possess apparent 2R hypothesis functionality: that capacity for genome duplication within an organism's evolutionary history. The notion was first proposed in 1970 by Susumu Ohno to better understand genetic latitude within vertebrates. The phrase was coined and situated, though Ohno's hypothesis[2] was rejected by Austin L. Hughes in 1999, writing in the *Journal of Molecular Evolution*.[3] Such debate enlivens and tests the legitimacy of pressing scientific progress, as it always has. But debate or no debate, humans are clearly outside the realm of any paleopolyploidy options, all our chips having been cashed, as it were, in an evolutionary casino a few millions of years ago. They no more afford the promise of radically beneficent genetic or morphological alteration. We are stuck, whereas *Branchiostoma lanceolatum* appears not to be.

Conceptuality is exempted from this apparent rule of no-exits. We can dream or dare to dream anything we like, as John Anster's 1835 translation of Goethe's *Faust* declared, in terms of boldness and genius. But our concepts of the biosphere have only devolved in the last thousands of years, following a pre-Paleolithic renaissance of openness to our interdependency upon, and (from most appearances) a certain joy in interacting with other species. Our sympatric cohorts who inhabit the same region, on every continent, are also members of our same species (notwithstanding our [certainly not universal] affinity with birds, dogs, and cats). And despite such human global distribution and unprecedented numerical multiplication, with estimates suggesting that at current fertility trends our species will easily surpass ten billion individuals, all of our options have been consolidated biologically into abstraction, conceptualization, thought, not morphology or novel genomes. If there is some evolutionary benefit to such mental *density*, it must emerge on ethical and artistic horizons, such as the Jains have so nobly and consistently enshrined within their daily routines, an ethical edge that somehow confers traction on our future survivability. Johann Sebastian Bach's 20 children by 2 wives suggests a most *creative* effort to propagate music, but one that in the twenty-first century would be deemed both irresponsible if not completely counter-collective and certainly unsustainable. So, as Malthus most famously argued, mere duplication is no answer.

More interesting, and clearly needed, are individuals who – despite their species' genetic cul de sacs – are able to function independent of a population (genetic) bottleneck and to do so by engaging other species in ways that offer some statistical

---

[1] Spruyt N, Delarbre C, Gachelin G and Laudet V (1998) Complete sequence of the amphioxus (*Branchiostoma lanceolatum*) mitochondrial genome: relations to vertebrates. Nucleic Acids Res 26(13):3279–3285, https://www.ncbi.nlm.nih.gov/pmc/articles/PMC147690/pdf/263279.pdf. Accessed 2 June 2017. 1998 Oxford University Press

[2] Dr. Susumu Ohno (1970) Evolution by Gene Duplication, Springer, ISBN: 978-3-642-86,661-6 (Print) 978-3-642-86,659-3 (Online), https://link.springer.com/book/10.1007%2F978-3-642-86659-3. Accessed 2 June 2017

[3] Hughes AL (1999). Phylogenies of developmentally important proteins do not support the hypothesis of two rounds of genome duplication early in vertebrate history. J Mol Evol 48(5):565–76. doi:https://doi.org/10.1007/PL00006499. PMID 10198122

advantage to the overall biome in which such interactions occur. As we have earlier postulated in our book, *Anthrozoology: Embracing Co-Existence in the Anthropocene* (Springer, Switzerland 2016), the "reciprocity potential" is equal to the square root of compassion, times the square root of the population mean divided by the sum of the entire group, in this case, *Homo sapiens*. What exactly does that mean? Our intent was to concretize a concatenation of global, biosemiotic data evidencing no less than that all species contain this RP, suggesting that they have ways to communicate with one another, to mutual good, no matter how complex and varied the so-called multiplication tables – rapid accumulation of environmental circumstances – with which they must cope. But it is merely an "equation" and all equations, not least the $E = mc^{2's}$, of the world reference a theoretical province of the imagination, wave after wave of colliding black holes which, in this case, are the unexpected phenomena inherent to the study of ethology and genetics here on Earth.

Consider the dialectic, ongoing, that bifurcates the conspecific and heterospecific gametes – whether sperm or pollen – whose fertility-driven anatomies (diploid, haploid, and the resulting zygote, etc.) are focused on the future offspring of one or differing species, from social insects to various vegetables to vertebrates. The differentiation of species has always called into question a parallel difference among individuals: This comparison exposes the zealous legacy of Linnaeus to the most fascinating enigma of our times, namely, the alleged competition of gametes and a correlated assumption (held by most people with the exception of such minds as Paulus Potter, Jan Brueghel the Elder, Erasmus Darwin, Percy Shelley, Maurice Maeterlinck, Hugh Lofting, and the like) that different species are, fundamentally, *different*.

Our contention is that those differences are merely physical (and under a microscope almost identical). That in the sphere of actual intelligence, sensitivity, the myriad nuances of feeling and psychology, they are even closer in kind. The human world, steeped in every version of bias, compounded by the nearly 7.5 billion biased individuals, has injected meaning into species differences where such meaning is only measurable and qualified by our *own* yardsticks. As adduced earlier on, our ongoing soliloquies have redefined all of natural history. "Hamlet" by Shakespeare's monologues represents a benchmark for a biology which, in general, has been badgered and manhandled specifically by *Homo sapiens*. This is a problem for future students of survival to thoroughly contemplate, as complex and urgent as terraforming, but on a level that transcends nuts and bolts, leading directly to the contemplation of the soul and its role in formulating what Roy Morrison has trenchantly described in a book entitled *Sustainability Sutra: An Ecological Investigation* (Select Books, New York, 2017).

Our focus is upon the individual, not the species or subspecies levels of taxonomic varieties. That multiplicity gives us tens of millions of species, and even more subspecies, varieties, races, clades, and the like. Often murky distinctions have given rise to a robust philosophical divide regarding the very definition and relevance of the species category. On one side, evolutionary theorists look to Linnaeus' binomial nomenclature as a persistently reliable, ineffable quality of re-constructivist life forms as categorized in a system that makes entire sense. On the other end of the spectrum, deep lineage and ancestral similarities of organisms and their populations – more similar than disparate – challenge every notion of fitness, each barrier to genetic ties, crossovers, hybridizations, and distribution, all leading to a serious rethinking of how humanity has endeavored to allegedly stamp order upon seeming chaos (by *allegedly*, the authors obviously doubt any degree of said *order*, save upon the written page). So that any rational human must regard all post-Renaissance systematics with no little skepticism, as with Big History Thresholds: biodiversity is a stunning study in the uncountable, in fuzzy logic and the trespassing of every category, the logical outcome of philosopher Gilbert Ryle's famed "category mistake."

A rash of such missteps, contraries, and zoological haze are predicated upon fertility infrastructure, all those machinations making for Darwinian evolution, and with it mutations, punctuated equilibrium, biogeography as in the case of the famed finches of the Galapagos or honeycreepers of Hawaii; species, in other words, whose island isolation from other nearby populations has only barely and temporarily ensured their survival in the form of finches, different color, and beak sizes. For the 18 remaining honeycreeper species, descendants of Eurasian rosefinches, their adaptive radiation between the four main islands of Hawaii (in terms of avian evolution) shows a spectacular variety of beak types, though it is believed that the survivors of what once was probably nearly 50 species are themselves presently quite vulnerable to extinction, at least one-third of their species critically so, according to researchers at the University of York and Max Planck Institute, most notably by the work of Professor Michi Hofreiter and team.[4] The honeycreeper isolation may represent one of the most fascinating insights to natural selection but also places a tragic window on the perils of isolation wherein the individual has no power to halt the demise of its species. Should we extrapolate all subsequent rules and outcomes of natural history on the basis of Hawaiian avifauna (and co-dependent plants and invertebrates), then we must admit to very little time, a mere fragment of a window, and a probability target of the most limited mathematical range for survivability among remaining taxons.

It has been theorized that the rosefinches made their way from Asia to Hawaii sometime between 5 and 7 million years ago, during an *irruption*, in which the species sought out new territories, either in large, small, or quite personalized groups. But when we refer to "the species," we are really thinking of a few males and females. How many, science does not know. It could be a large flock, or several flocks undertaking a remarkable journey into the unknown, guided by stars, electromagnetism, the equivalent of radar, and tens of millions of years of learned behavior – safety in numbers; or simply a few dreams, a male and a female might have sought the greener pastures out in the Pacific. This is where theories of group fitness, kin altruism and selection, and the evolution of eusociality promoted by the likes of E. O. Wilson's sociobiological construct hit up against the

---

[4] "Scientists Determine Family Tree for Most-Endangered Bird Family in the World," October 20, 2011, Newsdesk, Newsroom of the Smithsonian, http://newsdesk.si.edu/releases/scientists-determine-family-tree-most-endangered-bird-family-world. Accessed 3 June 2017

possibility that evolution actually has zero impact on individuals, only upon groups, a blessing in disguise, as we read it.

This would mean that for the better or worse, individual behavior is utterly unfettered and demonstrably free of the heavier patterns inflicted by any-and-all group theories, population dynamics, and social norms. Continuing, such a biological pattern would define a Leonardo as ostensibly immune to evolution, whereas the populations of Tuscany were not. It raises a host of unsolvable riddles: Leonardo, an offspring of Tuscan generations between the villages of Anchiano and Vinci, nonetheless fronting his individualism to remain untouched, if only by a notch or two, by his historical cultural context and, hence, the underlying biological rudiments of life in his household, his village, his region. It is a peculiar concept because it is at once liberating and constraining. Whereas in some transcendental manner, evolution – like an ether – hovers above the whole mechanism by which individuals are born and die, come and go, dream and disappear, there remains that wondrous scientific dare: the idea that in any population x, there are always going to be a rare percentage of individuals (population y) who truly manage to escape evolution during their lifetime, manipulating around whatever genetic traits and constraints they are born with.

By the age of 42, a Reinhold Messner had soloed, without oxygen, all 8000 meter peaks. While at age 31, Alex Honnold had free soloed El Capitan's Freerider route in Yosemite National Park. In evolutionary terms, Messner and Honnold may be as yet unheralded carriers of some 24th set of chromosomes. Are such ascent impulses (and genius) unique to individuals like a Messner and Honnold (or the late Hermann Buhl) that are not subject to the laws of population? Do their gut instincts escape somehow the first law of thermodynamics and the very gravitas of their species' population dynamics at large? Somehow managing to avoid the burdensome baggage of others during one's lifetime, however short or long? A John Keats, for example, dead at 26. A Percy Shelley, deceased at 29.

To rephrase: Can the individual bypass the impediments which, every million or so years, define the third and final act for most species on Shakespeare's stage, which is all the world? Or is our power of thought delimited by biological agencies to an extreme, as emblemized by that moment in "Hamlet," which reads: "There are *more* things in heaven and earth, Horatio, *Than* are dreamt of in *your philosophy*."

Remarkable phylogenetic data based upon examination of enormous catalogues of DNA, fossil, and family tree evidence were linked together into a compelling set of relationships and insights in mapping the destiny of Eurasian rosefinches on the Hawaiian islands.[5] But here again, the irruption resulting in such transmigrations could not have predicted the fatal competition from a species that did not even exist yet, humans. There is no algorithm to clinch this equivocation.

While global distribution suggests great opportunism, genetically, as in the case of wandering albatross and ants, this was no option for honeycreepers, or New

---

[5] See Lerner H, Meyer M, James H, Hofreiter M and Fleischer R (2011) Multilocus Resolution of Phylogeny and Timescale in the Extant Adaptive Radiation of Hawaiian Honeycreepers. Curr Biol, 21. doi:10.1016/j.cub.2011.09.039

Zealand kakapo, whose competitive advantages – expending no energy for millions of years by walking and climbing rata trees, especially when in mast, etc., as opposed to having to work hard to fly their heavy bodies around – has acted only to undermine their last options, as their respective species groups diminish down to sizes where each individual is microchipped, named/numbered, and monitored by we humans. Their life in the wild as individuals is actually a fiction in that their privacy no longer exists. They are in a biological test tube. And clearly for those who want to believe in design and purpose as part of the evolutionary strategy, to believe in God, for that matter, they must come to acknowledge that such a God did not anticipate the arrival of human beings and their impact upon such innocents as honeycreepers and flightless parrots (kakapo).

In the case of *Homo sapiens*, similar avifauna-like cul de sacs – and a massive measure of scrutiny from within – have occurred in all of the locations to which we have spread, indicting the notion that vast distribution is any more successful than isolationism. Indeed, the distribution theory in the case of *H. sapiens* is clearly the worst of all possible worlds, given its radical factor of added despoliation wherever its populations tread. Humans are largely incapable of letting go: we must destroy. Little evidence purports to the contrary, limited to the half-dozen or so tribes mentioned throughout this treatise. There is simply no genetic panacea awaiting our kind. Only a few noble ideals manifested as statistically minute population sizes (often in geographically isolated locations) and by our provisioning for those other countless creatures (e.g., *E. coli*) who ride along with or inside us and surely have reason to be grateful to their (otherwise freakishly self-destructive) hominid hosts.

Hence a peculiar and most likely disastrous dilemma: The most successful primate biologically speaking, humans, represent a perfect case study in biological delusion, the precise inability to find reproductive partners outside our own peerage. A massive case of sexual indwelling that offers no possible clue to evasive tactics, the ability to circumvent the genetic cul de sac our alleged success has engendered. Think of human evolutionary history as some fantastic maze into which we have, in our current form of *sapiens* willy-nilly wandered 40 years in the desert but during an ecological duration of nearly 8000 generations. We have not yet gotten away from the shadows of Mount Sinai. In other words, our exile has resulted in a repetitive loop within the shadows of our missed compass reading. We keep circling back because we are truly lost. Our achievement – nearly 8 billion strong and growing by an additional billion every 11–12 years – has succeeded only to the extent that there exists a root exudation of a globally distributed cancer which, in all of our passing wisdom, we assume we can retrofix by generically technological means (a meaningless summons). But there is zero evidence we can fix the problem of our clear and present ephemerality: That of the shortest lived large vertebrate in the annals of zoology to impact so proportionately large a habitat, indeed, *all* habitat. This demonstrative is devastating, every continent falling beneath our sweet-talking, wholly self-involved sword.

Genetic Fitness

If survival of the fittest is completely mistaken, given what we know to be a bewildering proliferation of meanings for "fitness" beyond the mere genotypic catalyst for offspring, then what, if anything, replaces it? There has long emerged a welcoming vogue to argue in favor of altruism, co-symbiosis, bilateral biophilia, cause and effect sheltered in the supposed fair winds of interspecies gentleness. Many have come of age believing that kindness and the ability of humans to be, for lack of a better word, good, will nurture a new kind of human consciousness that is much like the global mind mantras that have swept the ecological literature like a new age virus. That's not to be cynical, but concerned. While we dream away, focusing upon communitarian Utopias, we are daily escalating the victims of our convenient truths, namely, so many billions of other animals that we assassinate, and rarely with only a single bullet to the head.

From a genetic perspective, such fanciful paradigms might suggest a convenient obviation of that galling onus upon the human individual, shifting one's gaze, instead, to all those microorganisms for which the individual is host and to that congeries of companion animals and plants ranging in the uncountable billions. The aggregate of humankind bolsters the survivability of something like 7–10 trillion nonhuman organisms in each individual person. There is a certain ethical ambiguity to this inchoate Ark inasmuch as it is involuntary. But it is an astonishing headline to realize that at least 10% of ourselves is *not* human. That 70% of ourselves comprises water. We truly don't understand what we are, that remaining piecemeal bit of pure humanity, so called, within its physiological hourglass, and – at the same time – we bear little evidence of demonstrating the least patience with, let alone pure love for, any other species. This is one reason Jain monks will not waste tens of millions of sperm, for example, by having sex or extirpating bacteria by washing oneself with soap.

Enter now the additional problem we have entertained, namely, that the reproductive barriers between species are fickle at best, misunderstood and/or irrelevant; that the nature of a human individual is so off the mark in terms of the last several thousand years of scientific definition that we are left breathless with new possibilities, while our senses are stubbornly dulled by habit and that our cultural heritage has served up ad nauseam the same pointers with respect to how to behave and in what to believe, at least with respect to every social contract, dietary prescription, and self-eminence. Our sole hubris has more bright lights than a Milky Way or Yellow Brick Road.

But this is not intended to be a merely misanthropic harangue. The biological stakes are far too great for indulgences and luxuries of lamentation. Our destiny is so immediately clouded as to call upon something other than faith or science or cynicism to save us. Common sense and compassion are two undeviating guides that have worked in the past and could work in the future. We assuredly hope so and are depending on this scenario to effect some measure of remaining dignity. And with it, honesty and praxis.

Theorists for over a century have argued that there is no balanced ecosystem theory that can hold up to randomness or to the unpredictable vicissitudes of punctuated equilibrium. That even the theoretically perfect climax forest is actually in flux. More important than that tumult is the very understanding within the human brain, and its subsequent organizations that insist upon distinguishing between species, and – with similar vigor – presuming to label all *individuals* as members of a *particular* species, subspecies, or taxon, or some other type of being. In our rage for order there are dozens of categories upon which zoologists and molecular biologists presume to label all individuals as members of some group, as if the differences between individuals, which must necessarily invoke their imaginations, does not brush up against the species barrier. All individuals and their diverse mentations, their feelings and other sensibilities, according to this argument, are essentially identical: all humans have 23 chromosomal pairs, the 23rd being devoted to male versus female. Chimpanzees have 48 chromosomes, and that's the end of the story, goes the philosophical paradigm. Never mind that one may be a Stalin, another, Fra Angelico.

Let us take a pause and step back a moment: What are we actually saying in the foregoing? That taxonomy in concert with the principles of biogeography have simplified the actual truth of miraculous being? That the differences among organisms far outweighs the classifications that link them in phylogenetic congruences whose underlying bias defines evolution with a stern eye toward humanity being its fundamental purpose at the end of the day? This it is not, of that we are quite certain.

Ancient Greek philosophers acknowledged a very tender concept, *ta koina*, shared things. Aristotle and Plato, and later John Locke, argued over the minutia of *ta koina*. But in the end, our end, it has come down to what we perceive as common sense applications of our empathic sensibility. From any number of applied sciences, there is great variance within the human population for this predilection.

Regardless of its discrepancies, the Scottish philosopher Thomas Reid (1710–1796) enshrined a unified notion of human kindness and endeavored to front its wisdom as the hallmark of the so-called Scottish School of Common Sense in his 1764 publication of *An Inquiry Into the Human Mind on the Principles of Common Sense*. What one must take away from that vast work of feeling and scholarship is Reid's emphasis on the perception of the young adult unbound, of the child's pure perception of reality.[6] That purity is replete with suasion, capable of reshaping the world. An apple falls from the Flower of Kew tree on a child's head (the tree outside the home of the young Isaac Newton north of London in Lincolnshire at a Manor in Woolsthorpe-by-Colsterworth, as it is called). A coconut on another's head, someone in Bora Bora, for example, who lived and died and aside from a few relatives, the world has never acknowledged. A person, like most of us.

---

[6] See *Selections from the Scottish Philosophy of Common Sense*, ed. by G. A. Johnston (1915), essays by Thomas Reid, Adam Ferguson, James Beattie, and Dugald Stewart (online version). See also Boas G (1957) Dominant themes of modern philosophy: a history. Ronald Press Co., New York, p 660

The child is caught out in the whorl of nature/nurture tumult. Then, a mere 300,000+ years ago, enters Homo sapiens. If you happen to be a coyote pup, your eyes closed for your first 2 weeks, your monogamous parents, your mother's milk, is all that matters. When you open your eyes, the world is your oyster, it would seem. Think of this section in the following way, paraphrasing the opening of Saul Bellow's wonderful novel, *The Adventures of Augie March* (Viking Press, New York, 1953): our genes need not be our destiny.

The Coyote Conundrum

In 1758, Linnaeus' 10th edition of *Systema Natura* recognized *Canis* as a genus – dog, from the Latin. But the coyote of North America, *Canis latrans*, is etymologically diffuse, *latrans* linguistically defined as between barking and roaring.[7] Linnaeus remarked upon *Canis lupus* (the wolf) and *Canis aureus aureus* (the golden jackal). But it was not until 1823 that *Canis latrans* was actually delineated, by the remarkable young naturalist Thomas Say (1787–1834), whose name also appends numerous other species among a variety of classes.[8] Say was simply relaying information gleaned from Meriwether Lewis' specific observations of May 5, 1805, and written in his journals. Say was standing at the very site 15 miles along the Missouri River from the Platte River while participating in Major Stephen Long's expedition.[9]

By 1949, zoologists would identify another 18 subspecies of the coyote, making for 19 in all.[10] They include a fascinating constellation of geographically defined *C. latrans*, from Panama and Salvador, Honduras, and Belize through Mexico to the Rio Grande, parts of California, the Rockies clear through to the south- and northeastern United States and across numerous provinces of eastern Canada, and northern Alaska.[11] The varieties of coyote should come as no surprise. After all, among dogs, there are an estimated "155 breeds" and "The dog genome contains 2.8 billion base pairs of DNA in 39 pairs of chromosomes. There are 19,000 protein-coding genes, most of them with close counterparts in other mammals, including humans."[12]

---

[7] See Page RDM, Holmes EC (2009) Molecular evolution: a phylogenetic approach. John Wiley & Sons. ISBN 978-1-4443-1336-9

[8] See T. Say 1823. *Canis latrans*. Pp. 168 in Account of an expedition from Pittsburgh to the Rocky Mountains, H. C. Carey and I. Lea, eds., Philadelphia, Pennsylvania

[9] Edwin J, Long, SH, Thomas S, John A (1823) Account of an expedition from Pittsburgh to the Rocky Mountains, performed in the years 1819 and '20. Longman, Hurst, Pees, Orre & Brown, London, pp. 168–174; see also Mussulman J (2004) "Thomas Say, Canis latrans." Discovering Lewis & Clark. Retrieved 15 Jan 2013

[10] http://www.eol.org/pages/328608/overview. Accessed 19 Nov 2016

[11] See Way, JG (2013) Taxonomic Implications of Morphological and Genetic Differences in Northeastern Coyotes (Coywolves) (*Canis latrans* × *C. lycaon*), Western Coyotes (*C. latrans*), and Eastern Wolves (*C. lycaon* or *C. lupus lycaon*). Can Field-Nat 127(1):1–16

[12] "The Dog Genome," © 2010 Sinauer Associates, Inc., http://www.macmillanlearning.com/catalog/static/whf/life9epreview/sample_chapters/LIFE9ECH17.pdf. Accessed 20 Nov 2016

Indeed, "The wolf, dingo, dog, coyote, and golden jackal all have 78 chromosomes arranged in 39 pairs. This allows them to hybridize freely (barring size or behavioral constraints) and produce fertile offspring." In other words, whether gray wolf or golden jackal, Ethiopian wolf or dhole, African wild dog or African golden wolf – they can all interbreed with coyotes.[13]

With its expanding range, the coyote is widely believed to have more successfully adapted to North America than any over mammal, with the exception of humans (though – and this is critically important – not accounting for human depreciation and depredation of its habitat), in addition to squirrels, genus *Sciurus*, 29 species each bearing, as defined by the Latin name a *shadow tail*.

Squirrels are massacred (notwithstanding the love affairs people have with them in places like the greens of UCLA or Central Park) but coyotes even more so. The pressure to kill coyotes is perplexingly ever increasing. Over 20 million coyotes were slaughtered by humans in the twentieth century, and that does not take into account the crossover lethal doses of secondary toxins; rodenticides purchased over-the-counter; lethal substances like commercially sold rat poison, brodifacoum, as one example (a vitamin K anticoagulant) which over the course of many days and nights impedes blood clotting in mammals (think 3.3 Holocausts); and currently an estimated 750,000 killed annually by ranchers and government wildlife authorities. You can be sure the mortality figures are much higher since no wildlife agencies are ever tasked with faithfully publishing their data – some publish nothing, amid an in-house grab bag of public relations bravado and outright loathing of transparency, goaded on by a socio-pathological demonizing of this remarkable species. In the first decade after World War II, at least 6.5 million coyotes were slaughtered in America.

Under such ecological stressors, all concurrent with this utterly irrational, murderous penchant in man, the coyote's matrix has morphed into urban and suburban human settings, a plotline based upon a most simple explanation: easy pickings in backyards, near kitchens. Yet, according to assessments published by the IUCN, coyote populations and densities are actually expanding, despite the human species' resolve to wipe this fabulous being off the map. Moreover, there is a dearth of reliable data on coyote population fragmentations, underscored by the fact coyote numbers appear robust across their entire geographical expanse,[14] with a population density estimated at between 2.0 and 0.2–0.4 per sq. km throughout North America.[15]

---

[13] See Wurster-Hill DH and Centerwall WR (1982) The interrelationships of chromosome banding patterns in canids, mustelids, hyena, and felids. Cytogenet Cell Genet 34: 178–192. Wayne RK, Leonard JA, Vila C (2006) Chapter 19: Genetic analysis of dog domestication. In Documenting domestication: new genetic and archaeological paradigms. University of California Press. pp. 279–295. *ISBN 9780520246386*. See also Wayne R, Ostrander, EA (1999) Origin, genetic diversity, and genome structure of the domestic dog. Bio Essays 21(3):247–57. doi:https://doi.org/10.1002/(SICI)1521-1878(199903)21:3<247::AID-BIES9>3.0.CO;2-Z. PMID 10333734

[14] Gese EM, Bekoff M, Andelt W, Carbyn L, Knowlton F (2008) *Canis latrans*. The IUCN Red List of Threatened Species 2008: e.T3745A10056342. https://doi.org/10.2305/IUCN.UK.2008.RLTS.T3745A10056342.en. Accessed 20 Nov 2016

[15] Macdonald D, Kays RW (2005) Introduction. In: Nowak RM Walker's Carnivores of the World, Johns Hopkins University Press, Baltimore, MD, p 97, by M. Bekoff 1977, Anim Behav 29(2),

Indians may have worshipped the coyote (whose name is as least as old as early Aztec nomenclature) but, in contemporary scientific reality, there is a fog of data surrounding *Canis latrans*. Theories abound as to the pressures motivating litter size, or lack of any litter; the role of solitary, pair, and pack behavior in rural versus urban settings; reoccupation of burrows and territories; and most saliently the density of populations in any given alleged territorial region. In one report from the state of North Carolina by all indications, a coyote litter size is unknown (e.g., between 2 and 12), and even a single coyote's range can vary from 1000 to 16,000 acres.[16] For all of our scientific hubris with respect to the longtime study of mid-sized, home-based omnivores (the aforementioned Thomas Say first documenting their nominate categorization in 1823), we know plenty little about this marvelous sprite of a "trickster" mammal.

Humans have nearly exterminated coyotes in at least seven US states. But it remains unclear how many actual coyotes, at any one time, are to be found in the Western Hemisphere, their home. An estimated 2000 coyotes languish insanely in zoos. In the wild, typically, their survival rate varies from nearly zero to 14–15 years. Most blurred of all is their rate and extent of North American hybridizations. Historically, those three sympatric species included the "three native species of the genus *Canis*, coyote (*C. latrans*), Mexican wolf (*C. lupus baileyi*) and red wolf (*C. rufus*)."[17] But the Mexican wolf has allegedly gone extinct, and the red wolf is critically endangered.

These gaps in data are vastly unsettling and not a little surprising. But the variables are also confounding. The second most successful mammal in North America remains an utter mystery, enshrouded in confusion, bias, and vitriol. What is clear, however, is that the 19 coyote subspecies, in addition to gray and red wolves and dogs with whom the coyote can interbreed, throws into question the utter latitude which presents as a possibility the life of all mammals, save for one, *Homo sapiens*, who has already driven to extinction all other fellow Homos:

Homo habilis, *Homo heidelbergensis*, *Homo rudolfensis*, *Homo erectus*, *Homo floresiensis*, *Homo neanderthalensis*, and *Homo naledi*. While humans biologically hoist the banner of annihilation (as Anne H. Ehrlich, Gerardo Ceballos, and Paul

---

May 1981; see also pp. 332–350, "An observational study of scent-marking in coyotes, *Canis latrans*," by Wells MC, Bekoff M, Knowlton FF (1972) Preliminary interpretations of coyote population mechanics with some management implications. J Wildl Manage 36:369–382, and L. C., 1972. See also University of Nebraska – Lincoln DigitalCommons@University of Nebraska – Lincoln Symposium Proceedings—"Coyotes in the Southwest: A Compendium of Our Knowledge," (1995) Wildlife Damage Management, Internet Center for April 1995 COYOTE POPULATION PROCESSES REVISITED Frederick F. Knowlton Denver Wildlife Research Center, Utah State University, Logan, UT Eric M. Gese Denver Wildlife Research Center, Utah State University, Logan, UT, April 1995

[16] http://www.ncwildlife.org/Portals/0/Learning/documents/Species/Fox_CoyotePopulationsReport.pdf. Accessed 21 Nov 2016

[17] "Hailer F, Leonard JA, Harpending H (2008) Hybridization among three native North American *Canis* Species in a region of natural sympatry. PLoS ONE 3(10):e3333. Published online 2008 Oct 8. doi: https://doi.org/10.1371/journal.pone.0003333 https://www.ncbi.nlm.nih.gov/pmc/articles/PMC2556088/. Accessed 21 Nov

R. Ehrlich more than amply evidenced in their 2017 book, *The Annihilation of Nature*) coyotes – not without equally hungry stomachs as man – are selectively driven, withheld, restrained unto the cusp of what is absolutely essential for their survival, approximately 1.3 pounds of food per day, gathered and/or seized from whatever is available, whether a frog or boysenberries. There is not a moment of excess, by most empirical accounts, although in some geographic quadrants, when hunting in packs they may take down a deer. That is a far cry from a slaughterhouse.

But to make matters even more confusing, coyotes understand their own demographics, much like antarctic Weddell seals and the brown marmorated stink bug (*Halyomorpha halys*), each possessed of a pellucid expertise that selects for an adroit exercise of fertility opportunism and/or restraint, as the situation demands. Like most species, coyote behavior is predicated upon population stabilization, at least when unpressured by humans. "About 95% of the time, only one female (the dominant or alpha) in a pack breeds. Other females, physiologically capable of breeding, are 'behaviorally sterile'," writes Dr. Robert L. Crabtree, president and founder of the Yellowstone Ecological Research Center.[18]

Unlike the coyote, humans have zero latitude for hybridizations. The cautionary tale of the coyote strikes at the very intersection of ecological peril and promise, of biological constraints that have all but doomed our 7.4 billion *Homo sapiens* to a life with no other sensate future than our own, within a finite genetic and psycholinguistic bubble that has incessantly strived to isolate itself, mostly through sheer propagation, but sometimes with good intentions, aspiring poetics, honest labor, that co-creative facility that has sustained children drawing butterflies prior to being gassed to death at the Terezin concentration camp. We have not been sparing on the subject of cruelty, and this fact alone must condemn that species complicit in its continuations. Yet we must not forget the goodness, now and then, spread more generously than nihilists want to admit. That healthy custom cannot be denied; it is felt with a feverish and universal piety and adrenalin event upon stepping into any Amazon, Empty Quarter, fine museum, or opera hall.

Moreover, young scientists are celebrating what they call "earth optimism," a mantra hailed during the past 4 years at "optimism" conferences held by the Smithsonian Environmental Research Center (SERC).

But *C. latrans* is sending us a message. The biophilial bonds between two otherwise agonistic species are mirrored in the social modeling that has configured both geographical bias (far denser populations of coyotes within cities) but also a profound survival shift as well, wherein the urban or suburban coyote's life expectancies have also increased more than fourfold over their wild brethren. This is both tragic and ironic. From the coyote's point of view, and given the species' multiplicity of hybridizations from Panama to Maine to Alaska, it is remarkable that coyotes have retained a shy curiosity about humans – despite the horrors meted out by our kind toward *Canis latrans* – suggesting that side of their genome which encompasses

---

[18] In a letter dated 6/21/12, Bozeman, Montana, http://www.predatordefense.org/docs/coyotes_letter_Dr_Crabtree_06-21-12.pdf. Accessed 21 Nov 2016

the human-dog relationship dating back at least 10,000 years. A similar trait of forgiveness appears to exist with brown bears, wolves, most of the 400+ psittacine, species, pigeons, and many marine mammals.

Coyotes, when pressed, have the ability to look matrimonially toward other species and subspecies. Humans have no such luck. One could argue, of course, that with nearly 7.5 billion humans, the geography of our phenotypes is so enormous as to guarantee a continued revivification of our being. But that is a philosophy without biological substance (or the least common sense in face of a demographic abyss). The varieties of human genetic expression give no evidence of our ability as a species to actually transcend a uniformity of anatomical needs, though – interestingly – not at the individual level. For example, we know that a human can survive (under great pain and stress notwithstanding) solitary confinement. Or remain contentedly within a small geographical region his/her entire life. We recognize that human nomadic experience is equally adept at the negotiation of customary migratory itineraries, that refugees, expatriates, and others can commence new lives on new continents. Our physiological adaptability in such instances is key to our global expansion.

Coyotes appear to restrict their comfort zones (territories) to a relatively diminutive expanse, that varies, as earlier referenced to between 1000 and 16,000 acres, or the sum total of very few square miles, although their combined range encompasses all of Central and North America, and there are indications that coyotes are beginning to move farther south toward the Bolivian Altiplano and the Amazon.

Approximately 11 known communication forms are known to humans with respect to coyotes, but this is a typical instance of a human absence of empathic empiricism, of research modalities confined to a confirmation bias that only seeks out the most obvious utilitarian acoustics, shorn of any biosemiospheric nuance. Clearly, in the coyote's rich song traditions and variety, there are György Ligeti's and Winston Churchill's at every turn, as has been documented in so many parallel epiphanies regarding prairie dogs long championed by Dr. Con Slobodchikoff and colleagues at North Arizona State University.[19]

But what is most chilling about the comparisons we might adduce between the coyote and humans – the two vertebrates who appear to be most vigorous in their successful adaptive proclivities – is, on the one hand, the coyote's pertinacity over several million years, and, conversely, humanity's neonate path toward collective self-destruction, and the equally appalling ruination of nearly all other vertebrate species who venture anywhere near us.

Why have humans embarked on this course of action, and do we have evidence that this is the condition of individuals, or only of the collective? Moreover, does evidentiary logic have a chance of getting past the illogic of humanity? Of speaking to those who by all appearances refuse to listen, or are incapable of doing so? These are some of the urgent queries before this generation. How can we separate out the two – species from individual – particularly in light of the fact this question means everything to our future?

---

[19] Voight BF, Kudaravalli S, Wen X, Pritchard JK (2006) A map of recent positive selection in the human genome. PLoS Biol 4(3): 0446–0458 7 Mar 2006

Latitude or Cul de Sac?

A spate of publications in the first decade of the twenty-first century has questioned the future of human evolution, if realistically there is any. We realize that so broad and sweeping an "if" conjures up intellectual sloppiness. But there is no lack of tension in the air, no geopolitical calm that might assuage some other scientific place of refuge off the known map.

Hence, those scores of researchers seemingly obsessed with genetic salvation have most focused upon mutational probabilities within geographical commingling populations, "positive selections" across the human genome[20] and "natural selection" driving "population differentiation."[21] Many have equated human "progress" with evolution, but this strikes as counterintuitive, given the I = PAT and *tragedy of the commons* scenarios of ecological overshoot. Peter Ward, in a *Scientific American* essay,[22] focuses on the apparent convergence of future human evolution with increasing complexity, as well as the temptation to technologically enter into a series of symbioses with machines (ceding a mobile, presumably *normal* human life to a digital and virtual one). The potential downside to this, Ward points out, is that "The major obstacle to genetic engineering in humans will be the sheer complexity of the genome. Genes usually perform more than one function; conversely, functions are usually encoded by more than one gene. Because of this property, known as pleiotropy, tinkering with one gene can have unintended consequences."[23] But, according to a 2011 study, there are, in fact, a mere 60 mutations that are known to be driving any form of evolutionary potential between humans and their offspring. That is a minute permutational sample of possible change. Even if one accounts for the 83 million + *net* births per year (as of 2016), the computations simply don't favor any radical new subspecies of the Hominidae family. Even some of the most startling mutations, such as that of the MYH (myosin heavy chain proteins) 16 gene, in primates other than humans, the changes which may indeed have completely altered our evolution, our small jaw muscles versus those of many other primates, our subsequent diets, our brain size, the proportion of food-derived energy diverted to the maintenance of that brain, all of these took millions of years to come about, not weeks or months and certainly not according to an invented algorithm.[24]

---

[20] Barreiro LB, Laval G, Quach H, Patin E, Quintana-Murci L (2008) Natural selection has driven population differentiation in modern humans. Nat Genet 40(3):340–345

[21] Peter Ward, What may become of *Homo sapiens,* Scientific American. https://www.scientificamerican.com/article/what-may-become-of-homo-sapiens/. Accessed 23 Nov 2016

[22] ibid

[23] Slobodchikoff CN, Perla BS, Verdolin JL (2009) Prairie dogs: communication and community in an animal society, Harvard University Press, Cambridge, MA

[24] "How Many Genetic Mutations Do I Have?" By Natalie Wolchover | June 16, 2011 11:02 am ET, Live Science, http://www.livescience.com/33347-mutants-average-human-60-genetic-mutations.html. Accessed 23 Nov 2016. See also "Quora," "What are some genetic mutations that give a distinct advantage in humans?" by Ariel Williams, 10 June 2015

In other words, there is little likelihood of a golden jackal or red wolf equivalent – of any hybrid – with whom to mate and pass on to future generations a successful new addition to the human genome. We are more than likely blocked from ever achieving a greater number than our 19,000 –20,000 human protein-coding genes or the estimated average of 70 trillion cells per individual if calculated according to the average weight of all humans, namely, 62.0 kg (136.7 lb). The adult brain, approximately 1300–1400 grams, or some 3 pounds, accounts for a mere 2% of our body weight (again, on average) and those weights and proportions, too, look unlikely to ever change. Granted, *ever* is an impossible word to work with. *Forever*, even more so.

In comparing future human evolutionary scenarios, based upon mitochondrial DNA lineages in the brains, particularly cerebellum sizes and complexities, between domesticated and wild dogs, and wolves from the Arctic to India, Scottie Westfall discovered vast grounds for refuting any easy generalizations.[25] Every aspect of the canid brains, from the brain stem, thalamus gland, cerebrum, and olfactory bulb, has been seen to vary both over time and geography by sometimes enormous percentages. Given the astonishing paucity of empirical comparisons, or of any baseline for hybrids of any taxonomic terrain, it is clear that the multitudinous extant, as well as all the extinct Canidae family members, has actually enjoyed levels of evolutionary freedom our species has been completely denied. Despite the sheer bulwark of genes as a by-product of our vast numbers of individuals, we can only suggest that the densities of our geographical expansiveness have eliminated the remaining prospect for that degree of isolationism that could, theoretically, have procured a successful heritable new trait or mutation that might catch on. In not one biome within any of the seven (or eight if we add Zealandia) continents do we see the slightest chance, it would appear, of speciation. This is an enormously chilling admission with implications that are deeply provocative. In essence, we are left to our own devices which, to date, have proved to be ecologically solipsistic and suicidal.

"But if populations aren't isolated, crossbreeding makes it much less likely for potentially significant mutations to become established in the gene pool – and that's exactly where we are now," Ian Tattersall has said. In other words, future human evolution "is dead."[26] Brent Kopenhaver, Jr., contrasts "several evolutionary hypotheses"[27] with a fascinating overview of the important 1988 study by H. Harpending and P. Draper.[28] Harpending and Draper had examined the! Kung Bushmen of the Kalahari Desert of West Africa and Mundurucu inhabitants of the Amazon, declaring that the rigors of the Kalahari "forces all members of the tribe to

---

[25] "The problem with the claims about brain size and dog domestication," Natural History, 5 July 2012. Accessed 23 Nov 2016

[26] "Future humans: four ways we may, or may not, evolve," by James Owen for National Geographic News, 24 Nov 2009. http://news.nationalgeographic.com/news/2009/11/091124-origin-of-species-150-darwin-human-evolution.html. Accessed 23 Nov 2016

[27] "The Psychopath: A New Subspecies of Homo Sapiens," Brent Kopenhaver Jr. Sott.net, Sun, 09 May 2010 10:06 UTC, https://www.sott.net/article/208242-The-Psychopath-A-New-Subspecies-of-Homo-Sapiens#. Accessed 23 Nov 2016

[28] Antisocial behavior and the other side of cultural evolution. 1988. In: Moffitt TE, Mednick SA (eds) Biological contributions to crime causation. Martinus Nijhoff, Dordrecht

contribute to their collective survival [and that] reliable reciprocation of altruistic acts is crucial. The !Kung form nuclear families, stable, long term relationships and thus have a strongly selective pressure on pro-social behavior. Reproductive success is thus dependent on consistent collective altruism." Alternatively, the Mundurucu have few if any food acquisition constraints. "In their tribe, the women do most of the farming while the men compete physically and politically for social status. They spend most of their time engaging in gossip, fighting, planning warfare, and complex rituals, occasionally hunting for meat which can be traded for sex with the women. Here, reproductive success is dependent on the male's ability to compete in a social hierarchy; he needs good verbal skills, fearlessness to fight other men in physical competitions, and the ability to deceive women about the potential resources he can offer her. This environment favors the expression of the antisocial trait, which can potentially explain the evolutionary origin of psychopathy." Two amazingly different human pathways and destinies.

Indeed, evil and psychosis have been debated as distinctive traits delineating successful adaptations in the future, although it is unlikely that a generic Dr. Strangelove, igniting a nuclear war, has much chance of passing down his/her genes. More likely, "…in crisis there is opportunity, and in apocalypse there can be metamorphosis… the next system humanity creates will be far more sophisticated, fair, and abundant than our current civilization," declares Ellie Zolfagharifard in the *Daily Mail*.[29]

While there may be some comfort in such predictions, clearly, the majority of the foregoing indicates rather abundantly that *Homo sapiens* have boxed themselves in by every conceivable cultural, ecological, and geographical compass point; self-impositions masquerading as biological success (the demographic runaway train) that – by the minute – escalates our genetic isolation from all other species and, hence, our inability to meaningfully embrace them in any manner that might help us to elude a genetic cul de sac. And yet, some argue, this very narrowing of options is likely to hasten our near demise and thereby invite potential 11th h (29th day eutrophic) mutations that might favor a more mature human species. We simply don't know yet. This debate has been apotheosized into the theoretical and remains lodged in an intellectual purgatory that, in fact, does no one any good, like rhetoric, or checkers.

If we are to proceed toward amelioration of all that is perplexing about the intersection between the one and the many, we need roots: collective life forms that provide potential mirror images of what is possible in an individual. We need, in other words, a reliable definition of the individual. A construct that designs, contemplates, and ethically chooses consistently among a sufficient majority of life forms. Such that it might rearrange the myriad chaoses into some rational and reliable story line whose pulses ring true throughout observed natural history and the many artifacts that have been the guiding objects of our deliberations for tens of thousands of years.

---

[29] 11 Sept 2014, http://www.dailymail.co.uk/sciencetech/article-2752166/Are-evolving-NEW-type-human-Different-species-evolved-2050-scientist-says.html#ixzz4QuQZxSTK. Accessed 23 Nov 2016

Chapter 5
The Individual Versus the Collective

The Vicissitudes of the Self

In the gastrovascular crevice of most of the 6100 known species of Anthozoa (Anthos = flower, Zoa = animal), a cnidarian phylum class of reef-building corals, is the all prevalent *Symbiodinium dinoflagellate*, a eukaryotic *Chromalveolata* superclade, previously thought to be algae, whose symbiotic relationship from within the coral polyp is one of the great metaphors on Earth for successful collaboration. A vast scientific literature adores this conceptual flower-like animal who arranges itself in cellular beatitudes to empower and sustain the oceanic miracles that comprise those ever dwindling coral reefs around the planet. But what is most astonishing about this microscopic vortex at the heart and soul of the marine Amazonias is the relationship between individuals and whole communities of individuals.

That camaraderie is emblematic of both the lexicographical and philosophical debate at the square root of whatever it is that differentiates at the individual and species levels, as well as defining the significance, or not, of that alleged difference.

Stony coral, the theme herewith, seems to have come into existence some 500 million years ago, during the late Cambrian period.[1] They were soloists, distinct individuals who – with enough external pressure and a few actual extinction level events – chose, in many cases, to work together. Every aspect of their evolution indicates a honing of the senses: precise timing, perfect choreography, anticipatory genius. We terrestrial vertebrates typically find them beautiful and arresting. Who knows what *they* all think of us ungainly snorkelers, sightseers who haven't a clue how to accomplish what coral reefs have mastered, or worse, industrial trawlers and thieves in the night who have only contempt for these trillions of canny cells, each one a kind of research fellow at a great university.

---

[1] http://www.globalreefproject.com/coral-reef-history.php. Accessed 29 Nov 2016

Studies of dinoflagellate sensitivities have long been observed. For example, both light-sensitive opsins – retinal photoreceptors found in both the Cnidaria and in vertebrates – respond quite intensively to light levels, as would be expected. Descriptions of individual cells that have "escaped an area" from potential predators coincide with other examined "tactile" sensitivities in the organism.[2]

It is well known that coral polyps are exceedingly vulnerable to "turgidity" in their ambient marine environment.[3] Ultimately, the brilliance of a coral reef – the build-up of half-billion years of individualist sensibilities – falls prey to global human disturbance, most obvious in terms of climate change. Put simply, as marine layer temperatures rise, both surface and at depths, zooxanthellae, a largely photosynthetic species that attaches to the coral, and upon which the corals mostly depend (the zooxanthellae help procure the coral's food), cease their predominately endosymbiotic relationships, and the corals begin to die. Herein, the most obvious proof that environmental circumstances *do impact* individuals, a crisis which, in turn, breaks down communities whole. While other subsequent opportunistic zooxanthellae species try to occupy the symbiotic niche with the coral – the adaptive bleaching hypothesis[4] – if successive attempts also fail, as global warming increases thermal stress, accompanied by rising acidification marine layer chemistry, the syndrome results in noticeable bleaching. The reef aggregate begins to starve and with it the vast number of other aquatic dependent fish and invertebrate species. The coral may be infested by predatory infectious bacteria (e.g., Vibrio *shiloi*), the equivalent of bark beetles destroying the immune systems of terrestrial forests. As the coral declines in health, it loses its color, becoming white. Bleachings are increasing throughout most of the world's reefs, as most humans by now more or less understand, a devastating blow to global biodiversity. These mass bleachings – many dozens of them – were first identified beginning in the late 1970s.[5]

The genius of coral reefs, like that of every social insect colony, forest, root affiliation, and pollination orchestra with all their players on a world stage, represents a multitude of neurological and cognitive infrastructures combining what, to our way

---

[2] Moldrup M, Garm A (2012) See spectral sensitivity of phototaxis in the dinoflagellate *Kryptoperidinium foliaceum* and their reaction to physical encounters. J Exp Biol 215:2342–2346. Published by The Company of Biologists Ltd. doi:10.1242/jeb.066886. http://jeb.biologists.org/content/jexbio/215/13/2342.full.pdf. Accessed 30 Nov 2016; see also Hoppenrath M, Saldarriaga JF, Hansen G, Daugbjerg N, Henriksen P (2007) Dinoflagellate eyespot types. Baldinia anauniensis gen. et sp. nov.: a 'new' dinoflagellate from Lake Tovel, N. Italy. Phycologia 46:86–108; Moestrup Ø, Hansen G, Daugbjerg N (2008) Studies on woloszynskioid dinoflagellates. III: On the ultrastructure and phylogeny of Borghiella dodgei gen. et sp. nov., a cold-water species from Lake Tovel, N. Italy, and on B. tenuissima comb. nov. [syn. Woloszynskia tenuissima]. Phycologia 47:54–78

[3] See Erftemeijer PL, Riegl B, Hoeksema BW, Todd PA (2012) Environmental impacts of dredging and other sediment disturbances on corals: a review PubMed.Gov. Mar Pollut Bull 64(9):1737–1765. doi: 10.1016/j.marpolbul.2012.05.008. Epub 7 June 2012, https://www.ncbi.nlm.nih.gov/pubmed/22682583. Accessed 30 Nov 2016

[4] Baker AC, Starger CJ, McClanahan TR, Glynn PW (2004) Corals' adaptive response to climate change. Nature 430:741

[5] Chumkiew S, Jaroensutasinee M, Jaroensutasinee K (2011) Impact of global warming on coral reefs. Walailak J Sci Technol 8(2):111–29; Huppert A, Stone L (1998) Chaos in the Pacific's coral reef bleaching cycle. Am Natt 152(3):447–459. doi:https://doi.org/10.1086/286181. PMID 18811451

of thinking as primates, are sensory perceptions manifested according to any number of infinite selves. Their biosemiotic pathways toward health and understanding also connote a primeval, pragmatic humility that cedes a certain something in favor of a greater something, a greater self. By *great* we mean more materially viable for longer periods of time. The genetics and chemistry of hundreds of trillions of beings that have lived and died all told must comport with that which we best summarize as *individuals*. It is simply a question of semantics to think of all these beings as a *community* of beings. In reality, they are individuals and fall firmly within the legal and philosophical realms of attributions and correlations, of inalienable rights; dignity; a past, present, and future; a soul; an unarrested liberty; and imagination. A wit and a wisdom. That élan vital about which Henri Bergson wrote so compellingly in his book *Creative Evolution* (1907).

Human societies, conversely – as viewed in a linear time-frame – have sporadically flourished during the course of no more than several tens of thousands of years. Total biological novitiates and ingénues, we have everything to learn from corals; at the very instant our collective behavior is destroying them. According to researchers examining Australia's Great Barrier Reef, that primordially successful realm "had suffered the worst coral die-off ever recorded after being bathed this year in warm waters that bleached and then weakened the coral. About two-thirds of the shallow-water coral on the reef's previously pristine, 430-mile northern stretch is dead…".[6]

Beethoven, Darwin, and Parmenides

With the virtual dismissal by most scientists of Ernst Haeckel's biogenetic law (1866), namely, that ontogeny recapitulates phylogeny, that the individual's embryological relationship to its species-wide adult forms has a mathematically guaranteed mirror image, every single assumption surrounding evolution, politics, economics, experimental embryology, psychology, and biology has become enshrouded in an intellectual fog, now over 150 years old. No individual could intellectually replicate the shocking turnabouts resulting in the United Kingdom's decision to quit the European Union, or the American Electoral College to receive an American dictator who would walk away brazenly and insanely from the Paris Climate Accord, etc. And while many have seen the 2016 revisions of China's Dietary recommendations coming – suggesting that nearly 1.4 billion Chinese adopt consumption patterns limiting their eating of animal products by half, to fight climate change – no individual Chinese has the slightest clue how to actually effect such a massive shift in collective consciousness.[7] Sadly, an individual's dinner table is not the same as a supermarket, as much as we might light to extrapolate.

---

[6] Innis M (2016) Great barrier reef hit by worst coral die-off on record, scientists say. http://www.nytimes.com/2016/11/29/world/australia/great-barrier-reef-coral-bleaching.html?_r=0. Accessed 29 Nov 2016

[7] Milman O, Leavenworth S (2016) China's plan to cut meat consumption by 50% cheered by climate campaigners. The Guardian, Beijing, https://www.theguardian.com/world/2016/jun/20/chinas-meat-consumption-climate-change. Accessed 24 Nov 2016

A thousand years ago, a small tribe of Todas in southern India convened and the entire community agreed to adopt vegetarianism, despite their being surrounded by a variety of other, meat-eating tribes. What prompted the change? How successful was it? Can we, today, infer lessons from the Toda's *noyim*, or rules of behavior that might enable the divination from individual choices to collective behaviors?

These are just a few of the upsets that have challenged the great divide separating individuals from that famed line in Thomas Gray's "Elegy Written in a Country Churchyard" – "Far from the madding crowd's ignoble strife/Their sober wishes never learned to stray," a topic with which Elias Canetti grappled in his incisive book, *Crowds and Power*.[8]

Like tunicate marine worms, and mice who share certain genetic parallels with humans, coyotes also represent a potent intersect of behavior, genes, and population distributions that provide insights into human social equivalents and possibly our very future. We have, previously, explored coyote genetic latitudes which *Homo sapiens* lack, a Canidae family deep lineage asset that has proved of immense importance for purposes of interbreeding, survival, and range expansion. In the case of mice, the massive numbers of experiments by human investigators upon them have produced a retrograde vision of "knockout mouse" individuals[9] and whole projected ecosystems involving genetic manipulation that might engender entire living systems that lack that which they *would have* contained but for perverse human meddling.

With the tunicate larvae, there are well-known reasons for rethinking recapitulation theories based upon the ancestral similarities between vertebrates and invertebrates in the guise of the notochords and discs interstitially lodged between vertebrae, for example – a crucial evolutionary reminder that phenotypic variations and their concluding works of expression in a live, contemporary being are no more or less behaviorally, let alone ethically, predictable than, say, every note in Beethoven's "Piano Concerto No. 5 in E-Flat Major, Opus 73." This celebratory work of great genius has its own exclusive and separate internal ontogeny, obviously: (1) what some would rightly characterize as a Creator, in this case Beethoven, and (2) what others would argue is but one consequence of the slow mechanism working through genotypes and populations that is natural selection.

In this case of Beethoven, if by way of evidence to be gathered in the process of formulating a forensic theory of artistic embryology, suppose we had but a single note of that Piano Concerto and chose to label it an *individual note*, as in an individual. From that greatly attenuated vantage point, we would never be able to predict or imagine the rest of the completed concerto, absent Beethoven himself. And even if we had Beethoven, and a Rudolf Serkin at the piano and a Leonard Bernstein conducting, and Archduke Rudolph to whom the concerto was dedicated sitting their listening with a bottle of Tokay, and – to take a step even further, word from Beethoven himself that he was contemplating a composition that would come to be known as the "Emperor Concerto" – even if we had all of that before us and, take it even farther, notes, let us assume, written by Beethoven himself and explaining how

---

[8] Masse und Macht, Claassen Publishers, Hamburg (1960)

[9] See https://www.genome.gov/12514551/. Accessed 19 Nov 2016

he had precisely intended to inculcate three movements, an Allegro in E-flat major, an Adagio in B major, and a Rondo in E-flat major: even armed with all that information, poised to formulate an artistic outcome in the year 1808 (the year prior to Beethoven commencing the composition), we would still be nowhere.

Such a predictive caesura would be a bit like puttering about the monastery garden of St. Thomas's Abbey in the town of Brno in the Czech Republic and from that botanical foray, coming up with the Abbot Gregor Mendel's "Versuche über Pflanzenhybriden" ("Experiments on Plant Hybridization") paper which he delivered to the Moravian Natural History Society in February and March 1865 pertaining to his 1:3 ratio of inherited traits of the 22 types of peas he bred, namely, the laws of segregation, and of independent assortment, or an attempt to imagine Thomas Hunt Morgan's insatiable appetite for experimenting on the breeding of white- and red-eyed flies (perhaps an innate connection to white- and black-eyed peas).

The point in such far-flung analogies is this: An individual, at the level of universally admired genius, or some anonymous purveyor of ideas and feelings, in either case, we have no true premonitory advantage over the *group* when it comes to devising the blueprint of an individual's behavior. Had Hitler not failed the entrance exams at Vienna's Academy of Fine Arts in 1907 or 1908[10]; or had Napoleon Bonaparte's Grande Armée not suffered horrific losses of over 400,000 French troops during their retreat from wintry Russia during October of 1812, historians would be at a total loss to actually predict how the face of Europe and of the human world might be different today.

Similarly, had the Toba supervolcano in today's Sumatra not erupted some 75,000 years ago[11] arguably killing off all but some 10,000 human breeding pairs throughout the world, we have no way of ecologically computing the demographic escalation of our kind that would have occurred and, quite logically, changed everything pertaining to humanity's Malthusian growth rates and corresponding exploitation of all other species. We might well have failed and gone rapidly extinct, within a matter of a few decades.

Our predictive analytics – "autoregressive moving average models and vector autoregression models"[12] – cannot adequately account for the multitudinous variables, or what has been called a "stochastic target problem" in which the data of one, in a "controlled process," is incapable of ascertaining with certainty the inevitable random constraints that necessarily alter that process unrecognizably and/or subtly enough to utterly transform any outcome.[13]

---

[10] David W (2009) Face of a monster: self-portrait of Hitler painted when he was just 21 revealed at auction. Daily Mail, Accessed 24 Nov 2016

[11] Choi CQ (2013) Supervolcano not to blame for humanity's near-extinction. Live Science Contributor, http://www.livescience.com/29130-toba-supervolcano-effects.html. Accessed 24 Nov 2016

[12] Box G, Jenkins GM, Reinsel GC (1994) Time series analysis: forecasting and control, 3rd edn. Prentice Hall. ISBN 0130607746

[13] See Mete HS, Touzi N (2002) Stochastic target problems, dynamic programming, and viscosity solutions. SIAM J. CONTROL OPTIM. SIAM J Control Optim. Soc Ind Appl Math 41(2):404–424, http://www.cmap.polytechnique.fr/~touzi/st99siam.pdf. Accessed 24 Nov 2016

The very definition of "stochastic" intimates anything but a zero-sum logic: that set of randomly distributed data subject to analysis, but never to exactitude with respect to predictions. Variations start at zero in any attempt to verify a stable condition or process, but ultimately, the reality of stochastics hinges on a probability which – regardless of how much we may think we know – is always a 50/50 chance.

This is the case no more so than in the known rules dictating probability distributions within the broadest Darwinian inputs throughout the life of an organism, a population, and an interdependent array of ecosystems loosely fueled by the engines of evolution.

Stepping back from evolution to immerse oneself in ancient Greek philosophy, such as Plato's Dialogue from his Middle Period, "the Parmenides," only furthers a generalization regarding "the one" whose ineffable truth was said to be self-evident, the outgrowth of a logical assertion predicated upon a mystical experience that suffuses all of nature into an ungraspable singularity. It does nothing to help understand why, for example, we should worry about endangered species, or our own destiny. Focusing upon the classical dialectic between the one and the many, however, does accomplish a particularly pressing issue. Writes Rickless, "At the conclusion of Parmenides' criticism of Socrates' suggestion that forms might be thoughts, Socrates tries a completely different tack: he suggests that forms are patterns set in nature (*paradeigmata*) and that partaking of a form amounts to being like it."[14]

Throughout the Holocene (the marine isotope stages [MIS] used in paleoclimatology to define the interglacial warming period of the last nearly ten thousand years of post-Pleistocene hominid evolution) humanity, such as it is enshrined as a positive, psychological, and emotional affiliation both inter- and infra-species specific, has in essence lived through a highly agitated period of debate comparable to the many uncertainties that must emerge from this hotbed of archetypal conceptualizations, the *partaking of a form*, of being part of nature. Hence the confusion, parody, and the ironic in the fake sublime phrase *back to nature*.

Every vertebrate, we would adduce, understands it – both partaking and going back – while rejecting these natural proclivities at every turn. Ideally, we can reference these sentiments in our arts and sciences. But practically, our histories are comprised largely in fear of the mortality such grasping half-verisimilitudes present to our everyday imagination, an imagination that lacks a lexicon and is different from, though parallel with our grandest dreams and aspirations. In other words, we have written human history in the throes of one vast constellation of reactions to nature, always for better, but usually for worse. There is no other topic of thought or conversation that absorbs us more than the solace of a shade tree, the scent of food, a night ripened by stars, the always novel sunrise, all-subsuming sunset, a coyly conducive breeze greeting famously our cheek,

---

[14] 4.5 The likeness regress 132c–133a, http://plato.stanford.edu/entries/plato-parmenides/. Accessed Stanford Encyclopedia of Philosophy, Plato's Parmenides, First published Fri Aug 17, 2007; substantive revision Thu Jul 30, 2015, Accessed 25 Nov 2016, Copyright © 2015 by Samuel Rickless

the steady melodies of crickets by 10 pm or so. Our entire evolution may be summarized by that same eternal present of sitting around a campfire sharing stories, or simply staring into the flames with our solitary thoughts.

Moreover, all this *nature* gives us everything, even while we commute to work in a mob that we dread and despise.

Despite each successive modernity, we are never able to outpace our shadow, to get ahead of the nature inside and all around us, a muted and transformed nature. All the genetic engineering and reengineering, nanotechnology, and biocomputing that have absorbed humanity's notions of what its myriad future prospects hold, medically, ecologically, economically, and aeronautically (e.g., the colonization of Mars), lag far behind even the most basic catch-up, when it comes to biosemiotics. We simply have not learned to listen to a seagull, a sea slug, an octopus, or a redwood. While, in every documented case, it can be easily argued that humanity is assiduously attempting to partake "of a form" such that our lives, our societies, our biological future "amounts to being like it,"[15] "it" referring to nature (nature herself, rearranged and ascertained) according to Parmenides' eight deductions or arguments for instituting perceptual similitudes, all these intellectual assays easily elude us. We are not stupid. Parmenides was not stupid. We are simply, and mostly, lazy. While Parmenides intimated in a succinct cipher the notion that nature is whole, is one, is inside everything that is, that lives, that has shape and, ultimately, the very form that is life on Earth, within his eighth "fragment," Parmenides expressed the enigmatic belief that "the mind reunifies Being, which senses had mistakenly divided into many things."[16]

"Many things." What does it mean?

Ancient Greek overviews of nature grant a complex conceptual unity to all of biodiversity and render human self-perception but one (very personal) apprehension of ourselves within this infinitude of equal shares, equal perceptions, equal gods atop Olympus, an exquisite, ineffable stake in being aligned with all Others. This was no "ecological self," as many attuned to the deep ecologies have thought of it. In ancient Greece, we were inside the biological map of the Earth, and all of our philosophies were nothing more than an urgent expression of our place in the Cosmos, arrayed like so many other hundreds of trillions of sentient beings interdependently situated, a priori, as members of a community of life. We had, and we surely deserve no privileges. Our consciousness is not some communal purpose advanced by the firmaments, dozen + dimensions, and space/time continua. If it were, then Nietzsche was not wrong but hopelessly inept in his view that God is dead. God, should we be vindicated as a species, is very much alive, targeting victims every second of every day.

All such ideas are enlivened by their universality when speaking of the Holocene and anthropocentric biospheres. These notions of universal human-dominated etiology are empowered outright, precepts by which we live and breathe and stake our modest claim to participation. They presuppose a direct connection between

---

[15] Ibid., Rickless

[16] By Nolletti © Copyright in Italy 2004, in other countries 2013 – webmaster Federico Adamoli, www.parmenides-of-elea.net – last revision: January, 15 2013. Accessed 25 Nov 2016

individual and population expressions, suggesting that by the sixth century BCE in Greece, common sense prescribed a sensibility that all is one and one is all.

But this may well be a fiction, a premise that what one person feels, experiences, and hopes for can be instrumental in shaping the dynamic which sways an entire population, as our species verges on 8 billion, with the very gravid potential for surpassing even 10 or perhaps 11 or 12 billion people. With so many of us yet to emerge, the relationship between the one and the many blurs, becoming more like a blind and grasping confusion that is far off from any quaint or conclusive holistic umbrella called nature. We know this to be the case based upon the copious tallies of destruction *Homo sapiens* have failed to address, showing little willpower when it comes to group decisions that favor altruism over selfishness and compassion as opposed to a species-wide cave in to so much confusion born by the breeding process, previously referenced – of some 353,000 human babies that come into existence every 24 h (as of 2016). Ancient Greek philosophy could not possibly cope with such a reality. Nor can twenty-first century civilizations. Nor, for that matter, the 1964 Wilderness Act, written in 1956 by Howard Zahniser, or Justice William O. Douglas' famed brief recognizing personhood in streams and riverbanks, fish, and trees (Sierra Club v. Morton, 405 U.S. 727 (1972). As regal as such declarations are, we remain at a near complete loss. Worldwide, the average legal set aside for wilderness areas, parks, reserves, and green corridors is less than 13% of overall terrestrial acreage. In the oceans, lakes, and rivers, such protective mechanisms comprise barely 3%.

One may be many, but is not *for* the many, nor vice versa. Philosophically absolved from moral involvement in the plurality of things, however much we may fancy some grand unity of life forms, our fickle evolutionary pathways defy ecological correctness. In positing the one, we are also forcefully acknowledging that there be no obligatory ethic or action incumbent on our lives. Our posits are vacuous. Accordingly, the brunt of our observations and connections fall short of our ideals. Herein lies the rubbish heap of modern history, slack edifices, worn-out euphemisms, phrases that have zero purchase, imaginary doctrines whose very meaninglessness outwits our feckless attempts at sobriety. In sum, a set of neurological responses to our very own human proliferation that cannot possibly cope with a rash of overpopulated souls. We are lost in this labyrinth of ourselves, and this bewilderment constitutes a lethal legacy in that it has continued with increasing vengeance to let down the Others while elevating the self, with gravest of consequences. We live ephemerally and wounded within that chasm that has separated our kind from all others, as if the entire human population has hid out in the Grand Canyon or the South Pole, so deliberately removed are we (collectively) from a rational embrace of biologically distributive justice across the planet. We don't know what such a Utopian distribution might mean or look like. How it could function. We are able to place people on the Moon but cannot configure the rudiments of planetary equality among even a smattering of known life forms, let alone within our own species. Knowledge of suffering has not diminished our collective collaboration in its perpetuation.

Ultimately, the disconnects between individuals and large groups (in primatology, typically groups exceeding 150 individuals) show no signs of amelioration in any sector, despite the rare admission from a self-conscious yet pluralistic philosopher like the Romanian/French Emil Cioran who spent much time in the Luxembourg Gardens, who went 7 years with essentially no sleep (according to his *New York Times* obituary), and who, in 1972 writing from his modest Left Bank apartment in Paris, declared, "I found out then [...] what it means to be carried by the wave without the faintest trace of conviction. [...] I am now immune to it" (referring to his war-time support of Romania's Right-Wing Iron Guard).[17]

Such articulated immunity is effortlessly commensurate with isolation and depression, communion with nothing, a belief system bereft of anything sacred or even compelling (though few could argue the case that Cioran was not utterly interesting), angst in the morning, bereavement by evening. Cioran was certainly obsessed with the *worst*, a condition of intellectual paroxysms that shocked by its condoning of Nazism. He emphasized the failure of the human condition, the human experiment, without distinguishing the sum of those parts from his own ethically dubious details: daily stances, backfires, and betrayals. What is fascinating is that this obsession was so eloquently expressed, time and again. Terror all night long. Certainty that there was no reason to breathe or eat or think. That the most lovely birdsong mocks us and there is a slow worm eating away at the heart, that all is lost amid so much useless chatter and goings-on. Cioran's investment in pithy probity exerts a kind of time bomb in every one of his sentences that is destined to go off, no matter how much we reject the appalling prospects he lays bare. He enthusiastically embraced the collapse of humanity, spelling out its endless futilities. But in so doing, by way of autobiographically psychiatric confessions, he also made clear that the individual can either mirror her/his species, try to atone for it, or struggle to rectify wrongs of the self and of the collective, a case for which was so passionately embroidered in Caroline Kennedy's edited tome, *A Patriot's Handbook*.[18]

Psychiatric Genetics

We move from the Left Bank of Paris to that human problematique in general. When we say *human*, we search for answers, and the questions pivot upon behavioral anomalies. As if whatever goes wrong is the best fodder for what might go right. This is key to understanding the limitations of biological fieldwork that gives no quarter to the many critical metaphysical and ethical issues inherent to the very

---

[17] Ornea Z (1972) Poporanismul, [Populism] Bucuresti, p 198

[18] See Kennedy C (2003) A patriot's handbook: songs, poems, stories and speeches celebrating the Land We Love, selected and introduced. Hyperion Books, New York, For a penetrating examination of Cioran's life and contradictions; see Bradatan's C essay The philosopher of failure: Emil Cioran's Heights of Despair. Los Angeles Review of Books, November 28, 2016, https://lareviewofbooks.org/article/philosopher-failure-emil-ciorans-heights-despair/#!. Accessed 16 Aug 2017

traipsing about which is key to the universal trespass scientists and their Ph.D. students have always taken for granted as something they have the right to do, for some greater good. Never forget that many species, particularly those among avians, finally went extinct because of the very rush by all those artistic and scientific luminaries to get the very last ones for study and showcasing, before it was too late. And when it was too late, the value of their specimen collections escalated. There is no need here to name names: persons and institutions.

These extinction and anomaly syndromes are born only by humans. We see them in no other species. And there are no more conclusive endophenotypic expressions in human psychological breakdowns than those associated with diathesis stress pertaining to the 5-hydroxytryptamine receptor 1B, accompanying violent behavior in humans, and predispositional suicide attempts.[19] In each of the 21 global databases for these agonist/antagonist behaviors and gene locations in humans, rats, and mice, the underlying basis of intense study concerns individual and family-tied suicides.[20]

Such studies contrast with speculation regarding an entire species' suicide, the case for which was posed by Edward O. Wilson in an essay entitled "Is Humanity Suicidal?"[21] Wrote Wilson, and we quote at length, "The human species is, in a word, an environmental abnormality. It is possible that intelligence in the wrong kind of species was foreordained to be a fatal combination for the biosphere… Perhaps a law of evolution is that intelligence usually extinguishes itself. This admittedly dour scenario is based on what can be termed the juggernaut theory of human nature, which holds that people are programmed by their genetic heritage to be so selfish that a sense of global responsibility will come too late. Individuals place themselves first, family second, tribe third and the rest of the world a distant fourth. … In its neglect of the rest of life, exemptionalism fails definitively… But the world is too complicated to be turned into a garden. There is no biological homeostat that can be worked by humanity; to believe otherwise is to risk reducing a large part of Earth to a wasteland."[22]

Some have speculated that killing oneself may be adaptive, evoking various evolutionary theories of suicide. How ironic. Denys de Catanzaro's "mathematical

---

[19] See Zouk H, McGirr A, Lebel V, Benkelfat C, Rouleau G, Turecki G (2007) The effect of genetic variation of the serotonin 1B receptor gene on impulsive aggressive behavior and suicide. Am J Med Genet B Neuropsychiatr Genet 144B(8):996–1002. doi.10.1002/ajmg.b.30521. PMID 17510950

[20] Suicidal behavior may run in families. by Elizabeth Landau, CNN, http://www.cnn.com/2009/HEALTH/03/24/suicide.hereditary.families/, Accessed 27 Nov 2016; see also Qin P (2003) The relationship of suicide risk to family history of suicide and psychiatric disorders. http://www.psychiatrictimes.com/articles/relationship-suicide-risk-family-history-suicide-and-psychiatric-disorders. Accessed 27 Nov 2016; see also Turckei G (2001) Suicidal behavior: is there a genetic predisposition? Bipolar Disord 3(6):335–349, by, 11,843,783, PubMed – indexed for MEDLINE]. https://www.ncbi.nlm.nih.gov/pubmed/11843783, Accessed 3 Nov 2016; see also Science Daily, https://www.sciencedaily.com/releases/2010/06/100615105249.htm, Accessed 3 Nov 2016; aca-Garcia et al. (2009) Nucleotide variation in central nervous system genes among male suicide attempters. Am J Med Genet Part B Neuropsychiat Genet. doi: https://doi.org/10.1002/ajmg.b.30975, Accessed 27 Nov 2016

[21] New York Times Magazine, 30 May 1993

[22] Ibid., Wilson Op EO (ed) http://www.mysterium.com/suicidal.html. Accessed 27 Nov 2016

model of self-preservation and self-destruction"[23] posits self-preservation as a function of an individual's perceived reproductive chances of perpetuating "inclusive fitness." This "reproductive potential" takes into account such values as "continued existence" and the "kinship" accruing from each genetically related coefficient, namely, other members, however distant, of one's family.

"Each year, up to 20 million people worldwide attempt to commit suicide, with about a million of these completing the act," writes Jesse Bering. "That's a significant minority of deaths—and near deaths—in our species. A huge conundrum of near infinite courageousness in the face, admittedly, of so many madnesses. And there is reason to be suspicious that nonhuman animal models (such as parasitized bumblebees, beached whales, leaping lemmings and grieving chimpanzees) are good analogues to human suicide."[24]

If consciousness extinguishes itself according to some as yet indecipherable, albeit minimalist, evolutionary stratagem, there are existing models by which to possibly divine that playbook. Bacteria of the bladder are known to sacrifice themselves, for example. According to researcher Scott J. Hultgren, "We discovered that cells that line the bladder have a built-in defense mechanism that kicks in when bacteria attach to them – they commit suicide and slough off." In mouse bladders, Hultgren adds, "This process of bladder cell elimination is thought to be a natural defense mechanism of the urinary tract."[25]

Other versions of this mechanism were deciphered in 2008 by a team led by Dr. Bill Hughes of the University of Leeds' Faculty of Biological Sciences. They examined 100 million years' worth of insect monogamy versus polygamy to determine "why insects evolved to put the interests of the colony over the individual" and determined that sociobiologist E. O. Wilson was wrong, and Darwin was correct; that invertebrate altruism passes along genes through monogamous strategies. In concert with biologist Bill Hamilton's 1964 evolutionary theory, Hughes and team recognized "that in every group studied ancestral females were found to be monogamous, providing the first evidence that kin selection is fundamental to the evolution of social insects."[26] By such altruistic behavior, we find that, in quintessence, self-sacrifice for a greater concept of goodness and biological richness, at probably every level of life, is the most plausible scenario for all behavior, sexual motives

---

[23] de Catanzaro D (1986) A mathematical model of evolutionary pressures regulating self-preservation and self-destruction. Suicide Life-Threat Behav 16:166–181. doi:10.1111/j.1943-278X.1986.tb00350.x, http://onlinelibrary.wiley.com/doi/10.1111/j.1943-278X.1986.tb00350.x/abstract. Accessed 27 Nov 1986

[24] Is killing yourself adaptive? That depends: An evolutionary theory about suicide. Sci Am 11 Oct 2010, https://blogs.scientificamerican.com/bering-in-mind/is-killing-yourself-adaptive-that-depends-an-evolutionary-theory-about-suicide/. Accessed 27 Nov 2016

[25] Interstitial Cystitis [IC] Support Group, https://www.dailystrength.org/group/interstitial-cystitis-ic/discussion/bacterial-suicide. Accessed 4 Nov 2016

[26] Science News Altruism in social insects is a family affair. 30 May 2008, University of Leeds; Hughes WOH, Oldroyd BP, Beekman M, Ratnieks FLW (2008) Ancestral monogamy shows kin selection is key to the evolution of eusociality. Science 320(5880):1213. doi: https://doi.org/10.1126/science.1156108; University of Leeds. Altruism in social insects is a family affair. ScienceDaily, 30 May 2008. www.sciencedaily.com/releases/2008/05/080529141329.htm, https://www.sciencedaily.com/releases/2008/05/080529141329.htm. Accessed 16 Aug 2017

being but a pragmatic subset of the much weightier ethical predispositions of a communitarian goal. Suicide, asceticism, all forms of solitary life may well conform to the highest aspirations of the community, certainly in ecosystem dynamics where individuals – theoretical or hypothetical – by all indications play so absolute a role in the vitality of habitat and the coordination of the many. The accrual and accretion of individual actions and intentions is key to Jain metaphysics and is also fundamental to biodiversity. So let us be clear: Altruism is a psychological tenet claimed by multiple disciplines, but none have so much a stake as that of the individual. It is personal, whereas kin, species, and other forms of enlarged, enumerated predilection have no face. This may well be the crucial bifurcation point in understanding the biological role of an individual that is wholly different from that of the group concussion that may or may not result. Mathematical probability theories attempt to record and account for these correlations and are important to any study of population genetics and ecological dynamics, from Carl Friedrich Gauss' number theories to the $1/2n$ probabilities of coin and card tosses.

In the end, it comes down, we suspect, to the individual who makes decisions and enters the flow and the fray, both, with intentionality that has a raison d'être throughout every biome. The religious and scientific importance of this concept cannot be dismissed or undermined.

Paul Gauguin attempted suicide following the creation of his masterpiece, *D'où Venons Nous/Que Sommes Nous/Où Allons Nous* – "Where do we come from? Who are we? Where are we going?" in 1897.[27] Six years later, he died in his home and is buried in Atuona, Hiva 'Oa in the Marquesas of French Polynesia. Gauguin's queries presuppose the royal "we" and therein pose that ultimate risk, namely, an individual's renunciation of we. This was not an antithetical stance toward the collective and its inordinate leanings but, rather, we think, an apotheosis of that humbling-most and genuine personage, a person without fanfare, modest in every respect, doing her/his best out there in the remote and quite islands of the Pacific. The "I" hangs fallow and sacred, loose and removed from any coherence that might lend every adjective, pronoun, and subject to one's being. But in absence of a collective, that ego – for all of its noble intentions and fine friends – remains vulnerable to every dialectical anxiety and dangling modifier.

Meaninglessness attaches herself to a like-minded void within one's own self-image, positing caveats such as 1 day is the same as all days and one sentient being is the same as all sentient beings. There is no collective consciousness, only individual consciousnesses and consciences, accounting for the so-called collective thrust. But if that 1 day is akin to time, that one being equivalent to energy and mass, then the Earth and enveloping cosmos is the space, and we have some other prospects that may well differ from the much ballyhooed space/time/continuum. Indeed, the individual or singular sentient personage may well make for an entirely new *concept* of the universe, one accessible from individual sensory, intuitive, subconscious, preconscious, dream-driving, or cognitive repertoire, however anatomically limited, personal and specific. By defying physics not so much with biology as a general precept, but with the individual, we have arrived at a theoretical premise.

---

[27] Oil on Canvas, 139 × 375 cm, Museum of fine arts Boston, Public Domain

Nevertheless, we are still circumambulating that *concept* and must reiterate: Is it a question lodged by the one or the many? Does it reside perennially in all philosophical discourse and if so is that a human phenomenon, solely? Because we will never be able to know whether fleas engage in deeply psychoanalytic analysis? Or whether the wild lion that raised the leopard cub sense or know that this is most unusual? Or is that alleged rarity but one more anecdote we humans have contrived to further build a largely fringe argument holding to the belief that biophilia truly is systemic throughout all life forms, when perhaps it is indeed 100% self-evidently so, or, conversely, one more example – and universally trenchant so – of our Aesopian imagination working overtime?

All of that said and endlessly debated, we would do well to forget all the theories of collective unconscious, of atavistic repositories and mythology in a world of Sartre's character, Antoine Roquentin, and his famous little fictional town of Bouville. There symbols are abhorred and mentally demolished as the world loses, piece by piece, all of its cogency in the form of life, until it is thoroughly deserted, inhabited solely by a man of zero consequence, ourselves. Without connection, the *we* perishes without notice, while the individual, the "I," ascends to a nauseating nowhere. Absent adjoining sentience, or even acknowledgment of the plenum of creatures inside us, we are living in a non-place, as nonbeings, where the solitaire poses an anti-Christ to every ecosystem, while simultaneously building up a universal library, a Noah's Ark of individualism. This bizarre turn leads down every dark alley and holds up no torch of hope or mindfulness. Even the prospect of intentionality has lost out on a future horizon because it cannot bear the responsibility, let alone the personal guilt, of having been born. It must by definition strive to fend off the personal at all costs. Such is the illogic attending upon the debate that carries no name, not even an equivalent in any other context or framework. The paradigm for this ethos, in other words, is nonexistent.

Nonetheless, logic dictates that within the solitaire remains a tenacious inventory of memories self-generated. Cabinets of curiosities needing no stimuli. A treasure chest of inner being, often thought of as resourcefulness, or attitude, of resiliency and self-motivation. All those character traits that seem directed by a self, regardless of judgment from without, or even playmates as a child, are as free-floating and without meaning as the motes one sees in certain hallowed light. "Hallowed" unto the self, of course, as this is an aesthetic consideration. Further on according to the train of thoughtless thought, un-bullied, fully independent, the anchorite holds forth against every cenobite and occupies his/her mind with resplendence in blindness or across the full brightness of day, however much it may actually constitute a void. He/she doesn't care, or even notice. Too busy indulging it. As removed from any concern as the aforementioned Archduke Rudolph's heart in the crypt of Saint Wenceslas Cathedral at Olomouc (in the province of Moravia, Czech Republic), in reference to music by Beethoven.

There need be no psychiatry attendant upon the pathological introvert, obviously. Every philosopher must at some point give vent to solitude as the last safe-haven in a world that has gotten unfriendly. The bitterness of the self, unleashed upon a silent world, is the splendor of one's own governance. Yet, being deaf and blind to the unceasing biological vestiges of hope is no antidote, cannot better an individual's chances of happiness, and renders even melancholy a kind of sham. The "I" can and must do better than that if the world is to survive. This species altruism, a credo of ethological virtue whose dominion emerges from every self, undermines any cravings for solitude in that it must be admitted that every flower is a monastery of other working and praying flowers, each single *cracking* cry of the Adelie penguin, the evolutionary success story of a telling chirp that grew at length into a splendid and functioning rookery and then a colony. The many avifauna definitions of rook; colonialism; nesting birds of adjoining, adjacent, or co-symbiotic species; and colonies in general will invite any number of assumed percentages and adages. The sciences are soft when it comes to sorting out what constitutes a few versus the many – a species versus a collection of species, a community of communities. Then come the far more serious speculations, as they pertain to biosemiospheric song and orchestral arrangement – whether of pelagic sea-faring birds, or humpback whales. Each attests to individual breeding and collective staying power, a passion, it must appear, magnificently abetted by thought and corresponding communication channels – in other words, the attesting to Others of profound interest – in the face of spooky challenges we all cannot but acknowledge, one way or another. The lure of getting out of ourselves and confronting the greatest of all opportunities: our home, this fragile Earth, looms as an antidote to any and all psychiatric genetics and scientific nomenclature debates. But, most notably, it presages as a sublime harbinger.

Reclaiming the Individual amid Speciation

"In a square yard of turf…" Charles Darwin wrote to Asa Gray, "I have counted 20 species belonging to 18 genera."[28]

Consider how history has multiplied rather than subtracted, how science has been so enthralled – every discipline of the sciences – with multiplicities of one kind or another. We love to speak in the millions, billions, and trillions while thinking little of the twos and threes and fours. At the number one, we all, each and every one, fall swoon to the underlying truth of ourselves. The secret of the one is never lost when it is the one involved. But there is no doubting the mesmerizing gaze into a starry night, with its infinitudes, the eerie romance of all those grains of fine sand in an hourglass or, more philosophically, slipping through our fingers. Such evocations have certainly been the case with all things numerical in kind, from biodiversity to astrogeophysics. While we have referred to the ancient dialogue concerning the one and the many, it was actually not until 1974, and officially in 1978, that a "grand unified theory" in elementary particle physics, based upon robust work in a myriad of mathematical disciplines, was finally coined and appropriated.[29]

---

[28] Letter to Asa Gray, 5 Sep 1857 –www.darwinproject.ac.uk/letter/DCP-LETT-2136.xml; http://www.stephenjaygould.org/library/darwin_gray.html. Accessed 5 Nov 2016

[29] See Georgi H, Glashow SL (1974) Unity of all elementary particle forces. Phys Rev Lett 32:438–441. Bibcode:1974PhRvL.0.32..438G. doi:https://doi.org/10.1103/PhysRevLett.32.438

Just as we have counted galaxies and clusters, so too we are potentially hardwired genetically to feel compelled by a singularity in whatever guise of a portrait or unison: Rembrandt's (and in his manner, Bach's) self-portraits. And particularly Gerard ter Borch's elderly woman, possibly his mother, seated beside a table with a Bible (most likely) opened upon it, enshrined in the fixity of a matronly melancholy, her eyes beholden to an eternity that has spoken to her, prepared her, forewarned her, the epic sagas of humanity composed before a richly, dark-mushroom-colored otherwise blank verse of a wall, Holland circa 1672.

Evocations coming from such portraits impart the shared epiphanies – all those factoids one can adduce from, say, the many books on James Joyce – A-to-Z; or in Dostoyevsky's trepidatious little foreword to his *Brothers Karamazov*.

So many artistic nuances and mathematical connections add up, or divisively subtract from one stunning revelation after another. Reaching eyes in stained glass and hallowed light that brings so much multistoried luminance to life across so many eerily gaunt and heart-dampening cathedrals, shadow-play amid lifeless stone brought uncannily to life, similitudes of feeling evoked in our own untouchable selves. All this a dramaturgical biology that somehow cannot and will not miss out on the wondrous inextricabilities that are one soul to another.

All those whose very being and circumstance somehow marks our own inward sensations with a stunning accuracy that wanders, longs, and cannot but feel the shared pains and pleasures over a topography of centuries and sensation know that they are part of a voyage. These are the individuals we are searching for.

Amid the layers and throngs of so many artifacts and odysseys, personages, and haunting stares, we catch ourselves spending entire lifetimes in search of that one great love or sensation. One feels it in Ireland's oldest known pub, the Brazen Head in Dublin (built in 1198); the face of Ötzi, found in the Alps near Bolzano, Italy, said to have lived from 3345 to 3300 BCE; the two skeletons holding hands, buried at the Chapel of St. Morrell in Leicestershire County, England, from 700 years ago; the cremated remains of a 3-year-old found along the Tanana River in central Alaska, circa 11,500 BCE; and most incredibly, approximately 50,000-year-old burials at various Zagros mountain caves at Shanidar (most notably, number 3) in which, argue some paleontologists, the internments were peaceful and purposeful, and included the laying of dried flowers in tiaras on the skull of at least one adult male. Given much later evidence from approximately 1500 years ago near Modena, Italy, in which archaeologists uncovered a couple looking into each other's eyes, and scores of burial sites across ancient Egypt in which friends and couples were buried together, we have ample evidence to suggest a quality of mind accompanying the physical realities of being a person involved in social rituals, a sociology of the dead that must translate into a future of human possibilities.

Expanding upon this epiphenomenon of interdependent ethical mores, we can adduce from the scientific passions and global migrations of an Alexander von Humboldt and Aimé Bonpland, for example, an anthropological passion: they were interested in people, in what individuals did, thought about, what their lives were like. And this was certainly the case with the obsessive aforementioned Asa Gray. Questing after the "botanical Holy Grail," he spent "40 years, the greater part of his productive life," obsessed with "the memory of a fragmentary, dried, incomplete specimen in a neglected herbarium cabinet in France… haunted… the assurance of its existence as a living plant and the hope of its rediscovery were with him constantly. A shy, evergreen ground cover with dainty, creamy-white flowers in early spring; cheerful, shiny, bright green leaves in summer; a winter coloring rich and rare…".[30]

Not separate, but one with Gray, those many precursors to James Watson and Francis Crick's discovery of DNA's double-helical structure included such scientists as Friedrich Miescher, Phoebus Levene, and Erwin Chargaff. It was Chargaff who would recognize that "almost all DNA – no matter what organism or tissue type it comes from – maintains certain properties, even as its composition varies."[31] Such "properties" have also proved to be the through-story of nearly every genetic theory of evolution, juxtaposing each and every individual within a community. Therein lies, of course, all that makes for sense and confusion, joy and sorrow, the endlessly profound relationships to which all life owes her infinite identities, open-ended.

That plurality is key to ecological speciation and natural selection and equally crucial to mutations, incompatibilities, reproductive isolationism, vertical and horizontal gene transfers, and so-called genomic islands.[32] Ultimately, plurality is precisely what it indicates: the many of ones.

All references in language stem from an intension that must predispose a communiqué. Simply put, "the biological species concept," in which "A species is defined as a group of individuals that, in nature, are able to mate and produce viable, fertile offspring,"[33] has conditioned us to believe certain principles. They may be right or wrong. Or it doesn't matter since, obviously, "right and wrong" are just more human precepts that may have nothing to do with the world. Biology is far

---

[30] See Jenkins CF (1942) Asa Gray and his Quest for Shortia galacifolia. Arnoldia 2(3 & 4):13–28. JSTOR 42953488; Taylor N (1996) 1001 questions answered about flowers. Mineola, NY: Dover Publications, pp 47–48. ISBN 0-486-29099-9; *Shortia galacifolia*. Lady Bird Johnson Wildflower Center. Retrieved 1 Jan 2015; Jenkins CF (1991) Asa Gray and his Quest for *Shortia galacifolia*: special fiftieth anniversary issue. Arnoldia 51(4):4–11. JSTOR 42955144; "The search of the Carolina mountains returning from his trip abroad, Gray reached home early in November, 1839, and immediately plunged into the task of completing the Flora of North America. Shortia, however, was always in his mind. It was Michaux's incomplete and misleading label 'Hautes montagnes de Carolinie' on the herbarium specimen in Paris that delayed for nearly forty years the satisfaction he was to have in holding in his hand a living plant." p 17, http://arnoldia.arboretum. harvard.edu/pdf/articles/1942-2--asa-gray-and-his-quest-for-shortia-galacifolia.pdf. Accessed 6 Nov 2016

[31] See Pray LA (2008) Discovery of DNA structure and function: Watson and Crick. Nat Edu 1(1):100, Citable by Nature Education, http://www.nature.com/scitable/topicpage/discovery-of-dna-structure-and-function-watson-397. Accessed 5 Nov 2016

[32] See Safran RJ, Nosil P (2012) Nature Education, Citation: Safran RJ, Nosil P (2012) Speciation: the origin of new species. Nat Edu Knowledge 3(10):17, Knowledge Project, http://www.nature.com/scitable/knowledge/library/speciation-the-origin-of-new-species-26230527, Accessed 5 Nov 2016

[33] Source: Boundless (2016) The biological species concept. Bound Biol. Retrieved 03 Nov 2016 from https://www.boundless.com/biology/textbooks/boundless-biology-textbook/evolution-and-the-origin-of-species-18/formation-of-new-species-125/the-biological-species-concept-500-11726/

beyond these simplicities, even if human ethics is centered by them. But our judgments are as persuasive as a neutrality situated anywhere in the universe, without lore, gain or loss, preference or dimension. A space and time vulnerable to energy but whose expression may or may not have as an infinitesimal incarnation some idea in the mind of a mouse on planet Earth. Such are the dilly-dallies of ethology as well as genetics, evolution, and futurism. In other words, on the tree of life, so called, we are rank amateurs, fitting arbiters of the wind. "A Cloud in Trousers," wrote the poet Vladimir Mayakovsky in 1914. We are that molded clay about which Philip Rhinelander wrote his lean and elegant book, *Is Man Incomprehensible to Man?* (W. H. Freeman, New York, 1973), that holds potential, not dismay.

But what we are is much more complex, of course, than mere reproduction, anxiety, fear, hunger, pain, frustration, or Edenic euphoria. *Life histories* are intrinsically the most critical bio-constituencies that can be ascribed to any individual and its myriads of relations to others. In a study of population genomics published in *Nature*, the co-researchers led by J. Romiguier postulate, "Genetic diversity is the amount of variation observed between DNA sequences from distinct individuals of a given species. This pivotal concept of population genetics has implications for species health, domestication, management and conservation… Our analysis demonstrates the influence of long-term life-history strategies on species response to short-term environmental perturbations, a result with immediate implications for conservation policies."[34]

All such policies are steeped in *individual* concerns, critiques, and daunting prospects for the community, wherein the one is aggrandized and dramatically morphed into the many. Consider how this vast discrepancy, call it by any name – gulf, dialectic, or conversation – affects environmental governance (all politics): "…despite continued complaint about the USEPA's supposed protection of 'individuals,' most ecological assessment endpoints use populations or communities as entities with organism-level attributes, many use community entities with community attributes, and an increasing number use population entities with population attributes."[35]

And it is not merely the conflicted policies of protection and remediation that derive from those same definitional *implications*. In navigating the crowd in search of the individual, the most convincing argument attaches to the portrait, in a painting, photograph, on a stage, across the evocations of any work of art or lyric, a portrait enshrined in every act of nurturance, commiseration and empathy, agape and attachment, in every specific relationship involving activated senses. The soul speaks when it is spoken to, the most thrilling of all such implications, such that biology is

---

[34] See Romiguier J et al. (2014) Comparative population genomics in animals uncovers the determinants of genetic diversity. Nature 515:261–263, 12 Nov 2014, doi:10.1038/nature13685, http://www.nature.com/nature/journal/v515/n7526/full/nature13685.html. Accessed 29 Nov 2016

[35] Suter II GW, Norton SB, Fairbrother A (2005) Individuals versus organisms versus populations in the definition of ecological assessment endpoints. Off Res Deviron, US Environ Protection Agency, Integr Environ Assess Manag 1(4):397–400, 2005, SETAC, https://training.fws.gov/resources/course-resources/pesticides/Risk%20Assessment/Suter%20et%20al_IEAM_Nov_2005.pdf. Accessed 3 Nov 2016

transformed at once into vivid *biographies* whose proliferation amounts to nothing less than something majestically far more than the sum of its parts (and hence the failure of all social contracts and political motivations and the chimes of deconstructionism from Aristotle to Rousseau), which is precisely where the history of any species fails ever so saliently to account for individualism and its infinite expressions.

The Primacy of Gauguin's Coda

Just as the species definition fails the individual, so too genes cannot address the mystery of that same personage. Writes William Revelle, "Genes do not code for thoughts, feelings or behavior but rather code for proteins that regulate and modulate biological systems. Although promising work has been done searching for the biological bases of individual differences it is possible to sketch out these bases only in the broadest of terms."[36] The biological content of an individual is different from that of the species, and this is not simply a matter of arithmetic or fogged zoological mirrors. The aggregate requires an orientation that must, at least conceptually, get beyond what has been described as "the biological notion of individual."[37]

What is remarkable about today's prospective individual is that this person is the same individual she/he/it was at the very origins of life. We all inculcate, to varying degrees, those "lost tribes of Tamaulipas." If anything has changed, it is the nature of the discourse that now absorbs us, a narrative line that has the sense of history charging its poignancies, generating theories and hypotheses, but always restricted to a human recourse. The question, then: Does *human* mean human species or a human being? If they are differentiated, in what (albeit a most simplistic unknown) critical ways? Indeed, what are the very questions to be asked that seminally address that stark difference? And if there are meaningful answers, who is listening – the species, or the individual? In that gulf of understanding lies the demise of the entire Linnaean architecture and reaffirmation of some radically unrealized somatotropin, or peptide growth hormone germane to every visualization of Arcadia, that Paradise so persuasively envisioned by the Velvet Brueghel of early-seventeenth-century Antwerp or the Sultan Muhammad of sixteenth-century Tabriz.

And finally, why does any proposed substantive differentiation even matter? Won't the collective do what it is going to do regardless of an individual's thinking, feeling, or behavior?

In 1997, in a remarkable contribution to thinking about species, John C. Avise and Kurt Wollenberg concluded that "the 'species problem' cannot be properly addressed from a phylogenetic perspective without reference to the fine-focus details of pedigrees and of lineage sorting processes at the microevolutionary

---

[36] See Kazdin AE (ed) (in press) Individual differences. An Entry Encycl Psychol, http://www.personality-project.org/revelle/publications/ids.html. Accessed 3 Nov 2016

[37] The Biological Notion of Individual, first published Thu Aug 9, 2007; substantive revision Sat 12 Jan 2013, http://plato.stanford.edu/entries/biology-individual/. Accessed 3 Nov 2016

scales..."[38] What are those scales, if the individual is to matter? While there arises in the mind a self-evident, indeed requisite answer to that, conjecturally there is a divide, a temporal component accounting for hereditary traits, the conscious and/or instinctive predilections leading to reproductive occasion (mutual choice or otherwise), and – most uncharacteristically – the loner who, for whatever reasons or lack of reasons, never mates, or cannot mate, but has lived and died meaningfully. Where do we look for such signatures and handprints? Do they add up to ecological meaning with which contemporary scientific and conservationist contention must be weighed in order to better serve any philosophy of the individual that can stand the testimony of ecosystems that popularly favor species and populations?

In his fascinating book, *Species: A History of the Idea*,[39] Wilkins discussed some 26 different concepts for defining *species*. "To remedy this terminological inflation, I have christened them the Autapomorphic species concept and the Phylogenetic Taxon species concept." Wilkins describes the former by referring to a "geographically constrained group of individuals with some unique apomorphous character" [the counter-distinguishing of an organism from fellow taxons that have in common a uniform ancestor, as opposed to ancestry] and, with respect to the latter, "the smallest diagnosable cluster of individual organisms within which there is a parental pattern of ancestry and descent…" With these two emphatically delineated patterns of systemic being (endemically agglomerated cellular clusters no different, one might conclude, from star clusters), we can better grasp the biological attenuation of a language approach to conceiving of an individual versus the family (or, again, galaxy), of the family versus the community of families, of the population that serves genetic robustness to constitute, ultimately, a species. Within this family tree of individuals, we narrow our microscopes to encompass the one versus the many, yet again, irrespective of the ontogenetic origins that lead the organism to maturity and, hence, to the possibility of reproduction – a sure enough parachute enabling the mother to land 70% or so of the time on her feet (and higher percentages for asexual leaps out of the imaginary plane). These myriad lives in one were not lost to the earliest explorers of lens craft, such as the Delft-raised Renaissance microbiologist whose innovations involving one of the earliest microscopes yielded his famed worlds-within-worlds, real-time visualizations of a host of unicellular organisms from local pond water, or of blood flowing through capillaries: blood with all that lives within it.

Despite the revolutions in optics and our ever-increasing cognizance of life forms within life forms, the significance of that chain of events still misses the child and – most importantly – the child's own imagination. Given our strong need to grasp what it is the child knows, and how rapidly her/his acquisition of morphological and conceptual attributes takes place, we would be as a species best served to listen to

---

[38] Phylogenetics and the origin of species – (allelic genealogies/gene trees/lineages/mitochondrial DNA/phylogeography). Avise J, Wollenberg K (1997) OpenBook (NAS Colloquium), Genetics and the Origin of Species: From Darwin to Molecular Biology 60 Years After Dobzhansky, https://www.nap.edu/read/5923/chapter/11#61 Accessed 3 Nov 2016

[39] John S (2009) Wilkins explored engrossing speculations he first published in an essay entitled Evolving Thoughts. University of California Press, Berkeley/Los Angeles/London. http://scienceblogs.com/evolvingthoughts/2006/10/01/a-list-of-26-species-concepts/. Accessed 3 Nov 2016

children, and to be governed, certainly in part, by the child in ourselves, whatever that means, multiplied inexorably. Ask any Tommy Stubbins, or any injured squirrel in Puddleby-on-the-Marsh.

There can be little doubt of the child's overwhelming significance to culture and to cultural evolution. Mammalian nurturance, in particular, remains focused on the parenting conundrum: To what end are we making babies, and where are we leading those many children? From an evolutionary perspective, the twenty-first century remains an *apex predator of children*, even while seriously bad data has witnessed some amelioration. According to the World Health Organization, "5.9 million children under age five died in 2015, 16,000 every day... Globally, under-five mortality rate has decreased by 53%, from an estimated rate of 91 deaths per 1000 live births in 1990 to 43 deaths per 1000 live births in 2015... About 19,000 fewer children died every day in 2015 than in 1990, the baseline year for measuring progress."[40] Of course, that's simply a medical purview. What of the philosophical? At present, "131.4 million" babies come into the world each year, while another "55.3 million" perish.[41] At some 4 births per second, 353,000 per day, 24/7, the concept that we may not *know* what we're doing, given the proportions, ratios, scope, and extent of the thoroughly inflictive Anthropocene, would indicate two primary story lines. The first of these: the counterintuitive nature of so much reproductive behavior and accompanying child and maternal mortality (as well as child and maternal morbidity, not to mention hunger, malnutrition and stunting [nutritional illiteracy]) as against any rational measure of evolutionary health and equanimity. The second: we stand witness to a veritable orgy of parturitions amid an ongoing outbreak of biological Holocausts engendered solely by *Homo sapiens*.

In crass (and entirely artificial) economic terms, the equivalents to this population explosion, as Paul and Anne Ehrlich called it in their book by the same title in 1990, were published in August 2017 by Robert Costanza and Ida Kubiszewski. They write, "Results show that the global value of ecosystem services can either decrease by USD $51 trillion/yr or increase by USD $30 trillion/yr, depending on which scenario we choose. The latter scenario adopts policies similar to those required to achieve the UN Sustainable Development Goals, which would greatly enhance ecosystem services, human wellbeing, and sustainability."[42]

Among such chilling global fertility quotients and values (the TFR or total fertility rate) unequivocally impacting upon biodiversity loss at a level that is the most stringently refined equivocation within any mathematical approach to the systematics discipline,[43] we come back to those haunting words of Paul Gauguin in 1887: "Where do we come from? Who are we? Where are we going?"

---

[40] Global Health Observatory (GHO) Data, Under-Five Mortality. http://www.who.int/gho/child_health/mortality/mortality_under_five_text/en/. Accessed 30 Nov 2016

[41] World Birth and Death Rates. http://www.ecology.com/birth-death-rates/. Accessed 30 Nov 2016

[42] See http://www.idakub.com/academics/wp-content/uploads/2017/02/2017_J_Kubiszewski_ESscenarios.pdf. Accessed 19 Aug 2017

[43] https://www.newsecuritybeat.org/2012/03/hotspots-population-growth-in-areas-of-high-biodiversity-2/. Accessed 19 Aug 2017

The Primacy of Gauguin's Coda

Two studies, 32 years apart, lent insight to the Gauguin conundrum. In his famed work, *In the Name of Sanity*, Lewis Mumford, reflecting on John Stuart Mill's observations "at the height of Victorian optimism," wrote, "For this society, which had learned the mechanical art of multiplication, had neglected the ethical art of division; hence, though the condition of the worker actually improved, it did not do so sufficiently to counteract his discontents or overcome his natural anxieties."[44] This is a far cry, with respect to fully accounting for members of a species, when it comes to delicately parting the Red Sea between individuals and populations, populations and whole species. Note, the second study, in their section entitled "Species and Speciation," Daniel R. Brooks and E. O. Wiley write, "In at least some cases we can distinguish between species and populations by the observation that species are individuated information and cohesion systems whereas populations are not.... Biological species share both historical continuity and cohesion."[45]

In other words, the abyss separating who we are and how science views us is enormous, with little rapprochement in sight. This gulf was masterfully evoked in many of the works of Nikos Kazantzakis, particularly when he speaks not of any specific admiration for human beings, but for the flame that consumes human beings.[46]

---

[44] Mumford L (1954) In the name of sanity. Harcourt Brace and Company, New York, p 47

[45] Brooks DR, Wiley EO (1986) Evolution as entropy: toward a unified theory of biology. The University of Chicago Press, Chicago/London, p 177

[46] See Karanikas A The everyman of Nikos Kazantzakis: a basic lecture delivered before a variety of audiences. Emeritus University of Illinois, Chicago, Hellenic Communication Service, LLC, http://www.helleniccomserve.com/everyman.html. Accessed 7 June 2017

Chapter 6
The Separation of Mind from Species

Ethical Cognition in an Age of Despair

Moral realism has come to dominate much of the discussion points among contemporary philosophical disciplines. This is not in absence of a vast literature of continuing anthropological and ecological input, but the result of so much bad news, a madman in the throes of epidemic egophrenia, seconds away, potentially, from starting a nuclear war; and a distinguished CNN anchor asking a former Director of (US) National Intelligence, James Clapper, who had just questioned the US President's "fitness" to hold office, *What can we do*? Biological dissipation points, unresolvable forks in the road of increasing human incoherence, and deeply long-term double-binds demanding immediate choices and solutions for which we appear to have none. Our ancient virus of irrationality underlies the majority of our juridical, theosophical, and community standards which are splintered geographically in so myriad a chaos as to lend capitulation to despair. Example: most people who live with a dog consider him/her a member of the family; others think of it as barbecued dog ribs for dinner. Views on capital punishment and the human fetus vary with as much internecine anguish or fanaticism as other topics concerning socialism and taxes. Nations throughout the world see their voting constituencies deeply divided on every issue that matters most: personal freedom, edible food, drinking water, clean air, a pleasant environment, peace, and a roof over one's head, not one iota of which can be predicted or equitably shared in a world of 7.5 billion ungainly, indifferent, callous, and largely carnivorous *Homo sapiens*.

No matter how many lessons we absorb, our individual choices remain in constant turmoil and transition, with a perpetual mortal tension, here fixated on the abyss of all that we have learned from our pasts, or there stymied and hesitant to move at the juncture of opposing viewpoints. Cowardice rains down upon the mass bifurcation which is that of a divided and dangerous humanity. Our countless paroxysms have devolved to superstition. You are hiking alone up a trail where a rumored number of bears are known to live. You have it on authority that they tend to hibernate several thousand feet higher up the mountain. But this is early March and there is the climate change-related onset of particularly warm weather. The creek is wildly flowing and daffodils already creeping through soil that weeks before was under snow. Suddenly, a wind picks up. You look behind, to all sides. You stop and begin to go over in your mind the scenario and how you might respond should a bear discover you on the trail. You remember that bears make trails and use them methodically, day after day, year after year. They also happen to be among the most savvy vertebrates in any terrestrial environment.

Who is the intruder? You should not be there and you imagine yourself frantically scrambling up the nearest large ponderosa pine, or over a jagged wooden fence to the left or the right? Run toward the river, or simply run, period. Everything is now doomed to this one scenario, over and over again in the unceasing apparitions swarming your mind, all defying the Gandhi's, Zen mystics and "let-it-be" philosophers with whom, ordinarily, you think yourself in familiar company.

Within minutes you turn around and get out. Probably a good choice, on many counts. The anthrozoological Tasaday method, gentle collaboration, seems least likely by way of other choices at this moment in time. Put bluntly, you are overcome with anxiety and the slightest whisper of a noise jolts your nervous system with a pounding voltage.

If we live our lives this way, afraid, spooked, it is a reasonable assumption that the decisions we live by will be those of a frightened or a tired man. Indeed, we won't even get out of bed, eventually. The trouble is we have to get out of bed. To paraphrase Samuel Beckett, I must go on. Even if I can't go on. Whatever fears haunt us in the guise of information, that repository of data beginning back in time at the epicenter of a single sub-atomic particle does not comport necessarily with our personal lives. They are parallel – our lives and the whorl of macro- and biomolecules swirling all around us. With enough evolutionary pressure and greener horizons to invite enticement, our ontology will give rise to every Utopia, doomed or otherwise. We love bears. We crave their very existence. Yet, as Lord Byron wrote, "Sorrow is knowledge: they who know the Most/Must mourn the deepest o'er the fatal truth/The Tree of Knowledge is not that of Life."[1] But possibly it is? And, in this case of the envisioned bear, instinct (both human and Ursus) is everything; a keenness to the wind, the season, the sound of the creek, the early arrival of parsley family members, with attendant Black Swallowtail butterflies. And so on.

If this seemingly simplistic parable seems utterly incommensurate with the complexities of the real world, then there are many real worlds. This particular black bear scenario happened around 2 p.m. this afternoon. Near our home. A daily trail. Our daily bread. As woven into the psyche of human history as anything could be. The conflation of fear and of flight. Or, alternatively, some other philosophical promise, infused with all the hope for peace and tranquility, tone poems of little likelihood in those dark, damp Bavarian forests of the Middle Ages, said to be teeming with every imaginable danger and giving rise to all the mythologies of the

---

[1] "Manfred," Act 1, Scene 1, Lines 10-13, John Murray Publishers, London, 1817, pp 7–8

big, bad wolf; of the Brothers Grimm and of remote sylvan castles in Wolfram von Eschenbach's early thirteenth century *Parzival*.

But we are no less nervous about the impervious dark woods in the early twenty-first century, and – for all of our incessant domestication of nature – as hopeful as we are afraid. The mind seeks comfort in company. There is nothing new about this. Comfort in a croissant. A dictionary. A flickering candle in that genre of introspection and holy light repeatedly relied upon by a Georges de la Tour. We sit upon a chair and know that Pharaohs, Marilyn Monroe, Leonardo, and Charlie Chaplin all did the same thing. That door handles and beds remain of human scale, invariants in the claim to an anatomy within vertebrate evolution that depends upon quotidian routines and real-life comforts, with their engineering cohorts and necessities.

Comfort in the company of bears takes these relations to a new level. In metaphysics is the company of the unseen. There may be not one chemical equation for intuition, but there is no general expression for logic, either. In other words, that person who we imagine to be ourselves is devoid of any overall principle. The sum total of her/his behavior can be described, but never explained. The unseen is usurped by that which is seen only to the extent of our credulity. Coming face to face with a bear out in those "primeval woods" (as brilliantly illuminated in Simon Schama's *Landscape and Memory* (Alfred A. Knopf, New York, 1995)] may be the ultimate proof, if proof is needed, that a momentous confrontation with a single bear is a resounding resolution to a hypothesis: in Nature the individual most assuredly exists, far from her/his species, alone in the moment. If you are so fortunate as to be part of that sudden equation (and survive to write up the numerical coefficients on a blackboard), then the species part of the amalgam becomes a mere afterthought.

Which leaves us at the end of the path: unclear. At the conclusion of his labyrinthine arguments, poignantly parsed in piecemeal lucidity, Richard W. Miller has penned, "In general, the study of moral conflict establishes the adequacy of modern morality." Fair enough. There are good people. And it follows that there are good personages among every species. Bears. Dandelions. Dragonflies. Redwoods. Which segues nicely to Miller's final statement, "The great problems arise from ourselves and our societies, not our standards. Perhaps we should stop fussing about morality, and try to be good."[2]

There is always good to be found in the individual, despite what Aldous Huxley so famously described in analyzing the methods of Hitler in his *Brave New World Revisited*.[3] Irish physicist John Tyndall (1820–1893) provided sufficient remarks in his postscript on Niagara Falls, to this effect when he referenced how his personal, self-sacrificing guide, Conroy, managed to save a landscape painter who had been nearly swept over the falls and was clinging to a rock in the thick of the rapids, while the entire community of Niagara, with all of their ropes and manpower, were not up

---

[2] Miller RW (1992) Moral differences: truth, justice and conscience in a world of conflict. Princeton University Press, Princeton, New Jersey, p 392

[3] Harper & Brothers, New York, 1958, pp 51–57

to the elaborate task at hand.[4] Let us take Conroy to be that quintessential Mahavira, Christ, Buddha figure at the heart of madness and make of him an iconic truth whose energies and agency must persuade the rest of us beyond darkness.

Like Aldous Huxley and Bertrand Russell, Paul R. Ehrlich has never been so certain of such essences, as when he discusses the genetic disadvantages among golden-mantled ground squirrels of sounding off alarm calls. "The free-rider genotype" (those who remain silent, diving into their burrows, calling no attention to themselves) "would be the most fit, out-reproducing the carriers of altruistic genes, and silent ground squirrels would take over the population."[5] The great philosophical maverick, Elias Canetti, took this position of biological irony even farther when he writes of the paranoiac Daniel Paul Schreber's *Memoirs of My Nervous Illness* (in a book translated from the original German), "God also had a weakness for the language of the Germans." That "God's chosen people in history… were, in order, "the old Jews, the old Persians, the 'Greco-Romans' and lastly the Germans."[6]

Deliberately, human nature supposes, surmises, invents, or resists invention, as played out most notably in politics, as well as the history and practice of aesthetics. Mark Roskill examines a most profound dialectic, in his chapter on "The Languages of Landscape," addressing the fact that Dutch Renaissance painters principally accepted the landscape as it was (as they perceived it), which might account for the fact that "a good number of Rembrandt's landscapes were left unsold."[7] Whereas Philippe de Loutherbourg's "Eidophusikon" of the 1780s was a best-seller. Here was the celebrated painter who was a veteran of Paris Opera productions, advertised "Various Limitations of Natural Phenomena, represented by Moving Pictures" as represented in a "box eight feet deep, with a vertical picture area of sixty square feet or less…" by which he presented to the highly roused public various spectacles, such as the "Effects of the Dawn" and "Noon [in] the Port of Tangier, with [a] distant View of the Rock of Gibraltar."[8] In his profound treatise on "barely imagined beings," Caspar Henderson quotes the thirteenth century mystic Meister Eckhart who said, "we cannot see the visible except with the invisible."[9]

In his love letter to his fiancé Estelle Oldham, William Faulkner hand-bound a lengthy newly written poem entitled "Vision in Spring" during the very months that T. S. Eliot was composing "The Waste Land." Faulkner, as Judith Sensibar points out, wrote six such handmade books prior to his death in 1962. This one was particularly

---

[4] Tyndall J (1898) Fragments of science: a series of detached essays, addresses, and reviews, vol 1, D. Appleton And Company, New York, p 204

[5] Ehrlich PR (2000) Human natures – genes, cultures, and the human prospect. Island Press/Shearwater Books, Washington D.C., p 40

[6] Canetti E (1962) Crowds and Power, Translated from the German by Carol Stewart, The Viking Press, New York, p 445

[7] Roskill M (1997) The Languages of Landscape. The Pennsylvania State University Press, University Park, Pennsylvania, p 78

[8] Ibid., pp 78, 82–83

[9] Henderson C. The Book of Barely Imagined Beings – a twenty-first century bestiary. University of Chicago Press, p 109

trenchant. For it is within this haunting volume that Eckhart's "invisible" may be said to be reincarnated, as "Orpheus": "He stands and sings./The twilight is severed with waters always falling,/And heavy with budded flowers that never die,/And a voice ever calling, ever calling."[10] Faulkner speaks at the end, in the poem entitled "April" of "reaching hands," and incites that moment when "the wind and sky bend down, and kiss/Its simple, cool whitely breathless face"[11] as if to liberate their love, whether human, flower, or other animal species. It is here where the poet confronts the solipsistic bias of centuries, certainly within the Cartesian genealogy as echoed by those psycholinguists and philosophers who have held to the following generality: "Because of our symbolic abilities, we humans have access to a novel higher-order representation system... a means of representing features of a world that no other creature experiences, the world of the abstract..." And in the same cadence admits to a particular dilemma, namely, the assertion that "We possess no brain regions specially adapted for handling the immense flood of experiences from this world... [and] The consequences are both marvelous and horrendous... Philosophers have long struggled with the problem of how we know that we are in a world populated with other minds."[12] As if to tacitly ignore the problem of Beethoven taking walks through the 19th District of Vienna, the Döbling, or Wienerwald, along the slopes of the lovely Kahlenberg Mount. We could attribute to these sojourns the composer's working out of all the details of his 6th Symphony ("The Pastoral"). And we could equally attribute to every bird or insect that flies, the working out of some other mentation, work of art – imagination, with all those solitary hours at their disposal, a Wandering Albatross quite literally circumnavigating the planet; a spider spinning her web, some great sculpture; the mysterious ladybird-beetle migration in the Rockies, converging with great enthusiasm in tiny rocky notches between adjacent valleys for reasons not entirely clear – the more than 6000 species of the Coccinellidae family worldwide exhibiting more enigmas than possibly any other non-true bugs, their very flight some kind of deeply inbred ecological speculation and anticipation. And so forth.

The "problem" of understanding is strictly dependent on one's willingness, or not, to be defined and reified by custom and the habit of self-indulgent millennia, whether emic (insider) or etic (outsider). In both instances, the research and conclusions invariably stem from a narcissism that gives no quarter to the Others, let alone a mass of *Hippodamia convergens* in the act of migrating or hibernating.

In rare instances, only, has Maeterlinck or Erasmus Darwin ordained the spiritual, cognitive, and artistic truths of other Individuals. Or the poet Rebecca Hey the moral of flowers, in her lovely book by that same title (1833). Conversely, in his chapter entitled "Strangled Thoughts" (in *The New Gods*, Translated by Richard Howard, New York, 1974), E. M. Cioran has written time and again of a systemic scenario that has separated itself inherently from life, on the condition of a servitude

---

[10] Sensibar JL (1984) Introduction. In: Vision in Spring, University of Texas Press, Austin, p 76
[11] Ibid., p 88
[12] Deacon TW (1997) The symbolic species – the co-evolution of language and the brain. W. W. Norton & Company, New York, p 423

that prefers perverse continuation over all other possibilities. A continuity without the life force; in the absence of anything remotely connected to the zoological fanfare which is our only true heritage. He writes, "I die, in practice, I am lacerated by every anxiety which opens an abyss between life and death./Animals, birds, insects have resolved everything long since. Why try to do better? Nature loathes originality, nature rejects, execrates man."[13]

These are troubling dead reckonings of consciousness that lead by torment and temptation toward some cliff that calls. That vertigo of indifference has long exerted suasion in the grandiose human drama of collisions and self-regard. By whatever conceits and chicaneries we might characterize and evaluate our evolutionary thrust in the mishmash schematics of zoology, we invariably come out quite convinced of our stereotypical triumphs, hieratic preening, despite Holocausts and revelations of so many other cruelties and disasters. Louis Althusser's apotheosis of Karl Marx's thought as "theoretical practice" touches upon the French philosopher's so-called epistemological break that is the governing metaphor in that inalienable separation between thought and reality, consciousness and the objects of consciousness, a sense of one-Self and the reality of the Self. Broken mirrors whose shards and clouded glass make up our species' singular history.

Beyond Self

Is there a rational, if sacrosanct scenario that plausibly engages the entire *samavasarana*, as Jain metaphysicians have called that gathering which encompasses all within a refuge?[14] Saint Francis endears those steeped in Western traditions to this proof positive collectivity of affirmative voices and souls searching together for goodness in one another's intimate company. We have described both the theorems and theories – a range of inferential rules, both conceptual and within the realms of praxis – for what we have termed the "Reciprocity Potential" emergent throughout the realm of anthrozoological ethics which, we hold, favorably persuades addicts of conventional evolution to an altruistic orientation which, they believe, cannot and should not ever be reversed or even questioned in the large sense.[15] It should come as no surprise that in his "Session of March 4, 1978," at the College de France, 2 years

---

[13] Cioran EM (1969) The New Gods, Translated from the French by Richard Howard, Quadrangle/The New York Times Book Co., New York, p 102

[14] See http://www.tattvagyan.com/24-tirthankars/shri-adinath-bhagwan/description-samavasarana/; see also http://www.metmuseum.org/art/collection/search/74993

[15] See Tobias MC, Morrison JG (2017) Anthrozoology: Embracing Co-Existence in the Anthropocene. Springer International Publishing AG, Cham, Switzerland, p 320; see also Jacques Derrida's 1997 10-hour address, "The Autobiographical Animal," in *The Animal That Therefore I Am*, edited by Marie-Louise Mallet, Translated by David Wills, Fordham University Press, New York, 2008

prior to his death, Roland Barthes equated "individuation" with the most compelling of outward modalities, namely, "sweetness" and "gentleness."[16]

But these qualities of mind are strict deviations from the so-called median of probability that dictates the rubrics of extreme value theory, in the manner of generalized Pareto (continuous) distributions. Again, like Faulkner to his bride to be, and the omniscient Barthes who in February of 1980 would be struck down by a laundry truck while he was walking home on a street in Paris, and die one month later at the age of 64, so too in the Getty Museum's most important illuminated Renaissance manuscript, the Gualenghi-d'Este Hours, Bernard of Clairvaux's Sermons on the Song of Songs elicits "the feet as the appropriate starting point for 'the beginner on the way to salvation': Eyes that are accustomed only to darkness will be dazzled by the brightness of the spiritual world, overpowered by its splendor."[17] As for the ontology of laundry trucks, that is a separate issue.

Such revelatory immersions – the precocious awareness of that "splendor" – overpower posterity and give wakefulness and intention to the present predicament of the Anthropocene epoch, with startling clarity: beyond ourselves is the world of life, and the future belongs to each of those Sentient Beings who accept humbly their mortgage on this existence from the biomes of their primordial birth. The fascinating and relevant questions this connection beyond the Self imposes reach a critical mass when translated into such fundamentals as humanity's relationship through its neurological corridors to the sky and the oceans, the two most omnipresent facts surrounding us, the bulwarks beyond statistical compare comprising this one home we know, the Earth. Asks Michel Pastoureau in his accessible journey entitled *Blue – The History of a Color,* "The notion of a favorite color is itself extremely fluid. Can you say in absolute terms, outside of any context, what color you prefer?"[18] The relevancy of this seemingly nondistinct query hinges on more than mere culturally different predilections (30% of those Japanese polled on the subject, says Pastoureau, generally lean toward white, for example, whereas children across the planet apparently "prefer red," "always and everywhere…with either yellow or blue in second place."[19] We might theorize all day upon the significance of these (albeit, limited surveys) but will arrive at no sound conclusion, either in the realms of sociobiology or evolutionary group fitness theories.

What remains vulnerable to conclusiveness amidst such disparate minutia is some semblance of the fragmented arguments inexorably rooting human experience to the Earth and the elements, predisposing all perceptual limits, judgment, and

---

[16] See *The Neutral,* by Roland Barthes, Lecture Course at the College de France (1977–1978), translated by Rosalind E. Krauss and Denis Hollier, text established, annotated, and presented by Thomas Clerc under the direction of Eric Marty, Columbia University Press, New York, 2005, p 36

[17] Bernard of Clairvaux, *On the Song of Songs,* vol. 1, Translated by K. Walsh, O.C.S.O., Kalamazoo, 1981, quoted in *The Gualenghi-d'Este Hours – Art and Devotion in Renaissance Ferrara,* by Kurt Barstow, The J. Paul Getty Museum, Los Angeles, 2000, p 191

[18] Translated from the French by Markus I. Cruse, Princeton University Press, Princeton New Jersey, 2001, p 171

[19] Ibid., p 170

choice to biology. That being said, whether human Saint, a Michelangelo upside down for 3 years working away at the Sistine Chapel, or a busy mosquito somewhere right now in the Brooks Range of Alaska, each and every one of us (and by us we refer to Everyone – every living being on the planet) there is a profound Individual whose grasp of this fleeting life connotes nothing less than a presumed Right to be here, a need ordained by hundreds of millions of years of such rights, and an equally profound connection that represents the biospheric richness, diversity, and simulacra as a priority like no other in the cold calculus of the Cosmos which we observe with ever-increasing interest and precision. We do so with sophistical tools that, for all of their engineering marvels, day by day, have yet to equip our hearts and souls with anything more useful than those aforementioned words of Saint Bernard. The Earth gives rise to the future individual, ensuring a new nature in every one of us, were we to be open to the experience and its inestimable probabilities.

By "Self" we mean to suggest all that is indivisible, conjoined in the immutable, thoughtful without Ego, humble by nature. Enthralled, imaginative, aware. A singularity in the sense of an individual whose connections are not to a species, only (as defined emphatically by Linnaeus in the 10th edition of his masterpiece, *Systema Natura*, in 1758), but to every personage.

Biocomputing the "Personage"

One of the most recent and thorough overviews of the biological discussion of individualism, outside the realm of any theories of mind (like those of an Althusser), comes from a comprehensive essay, "The Biological Notion of Individual."[20] First published nearly a decade ago (August 9, 2007, and revised substantially since that time), the Authors have comprehensively discussed a myriad of seminal crossroads that must engage anyone focused upon the quintessence of "Biological Individuals." It is a brilliant overview, essential reading. Our intention is only to highlight a few of the most, for our purposes, salient dilemmas which Wilson and Barker pose.

For example, Part 5, "The Tripartite View of Organisms and Homeostatic Property Cluster Kinds." There they write, "In summary, the Tripartite View holds that any organism is physically continuous and bounded and is: a. a living thing (individual, agent) during at least some of its existence, b. that belongs to a reproductive lineage, some of whose members have the potential to possess an intergenerational life cycle, and c. which has minimal functional autonomy of the relevant kind." They define a "living agent" as an HPC, which refers to "Homeostatic Property Cluster," a rather complicated definition which trades between semantics,

---

[20] Wilson, Robert A. and Barker, Matthew, "The Biological Notion of Individual," *The Stanford Encyclopedia of Philosophy* (Spring 2017 Edition), Edward N. Zalta (ed.), forthcoming URL = <https://plato.stanford.edu/archives/spr2017/entries/biology-individual/>. Accessed 8 Mar 2017

empirical observation, philosophical abstractions, and human psychiatric modalities all questioning whether nature or humans have monopolized the endless sets of characteristics which defend or hold harmless the clustering of biological entities and the terms used to reference them as connected, associated, disconnected, contiguous, separate and apart, or holistic and self-referential.[21]

Other crucial Wilson/Barker points of interest concern the enumeration of 20 types of individuals, including all of the conventional evolutionary organisms and system relations, but also "first/last living thing," sterile creatures, "lineage-based, biological individuals that are not living things" (an essentially Daoist consideration of rocks, a Jain deliberation upon dew drops), artificial life forms, and all of the great collectives (e.g., coral reefs). "Groups," "Trait Groups," "Superorganisms," "Clades," "groups 'above' and genes 'below,'" models of biological "pluralism," and the contention that "colony-level, group, individual, and kin selection are all aspects of gene selection." This leads to their ultimate question: "In what ways does the study of genetic variation within our species constrain or even dictate how we should think about human variation more generally?"[22]

Amidst or Among?

Such nature/nurture predilections, genomes and mutations, traits, and mathematical models of heritability invite consideration of basic distinction: amidst or among. All cohabit the biosemiosphere, but to what extent they communicate with one another, are mutually dependent, and enjoy each other's company may well be a question of *mind* as much as it is of species differentiation. The two *objects* of our focus are clearly pivotal, congeal, and harmonize in ways far beyond our ken. We know it to be "far" because we cannot see it. Our intuition takes precedence over logic. The Latin verb, *intueri*, refers to contemplation. But its joint ineffability – a passive verb, an intransitive verb, the ongoing lack of a signature – harbors no solutions, as modern psychology insists upon, leaving the sojourner lost in a monastery of the mind, as John Keats likened a similar condition of human perplexity.

Theories of mind have long abounded, but they afford little or no comfort with respect to the independent Self. Psycholinguistics offers little balm in this regard. Noam Chomsky's "Universal Grammar" can be traced back to the mid-thirteenth century Modistae, philosopher/linguists like Martin of Dacia who had focused upon three components said to inhere in all humans, the first being that of *modi essendi* (modes of being), the other two comprising modes of understanding and of signifying. Variations were long accepted as crucial aspects of what would, many centuries

---

[21] See https://georgschauer.com/2012/12/10/the-causal-mechanism-in-boyds-homeostatic-property-cluster-kind-approach/. Accessed 8 Mar 2017

[22] Lewontin RC (1982) Human diversity. Scientific American, New York. See also Lewontin's book by the same title, *Scientific American Library Series* (Book 3), W.H. Freeman & Company, San Francisco, 1995

later, be viewed as genetic predispositions to the universal rules of language. No person was capable of acquiring more than a smattering of languages and, within a local language, an equally constrained vocabulary. It was thought that such natural impedimenta allowed for the acquisition of at least one's native tongue.

The singularity problem of human linguistics (most people speak only one language) must embrace psycholinguistics as the most plausible departure point for an insularity chiefly responsible for the entire life-span of a person within her/his clan. Yet, in command of a single language, that person and his nearest relations (a family sharing the same igloo) communicate both essentials and song; humor and grief, obeying intangible rules of a universal deep logic (some have suggested "deep mind") that has swept over occupation of every hamlet by humans anywhere in the world. That coefficient of the evolving human brain has, to date, been isolated morphologically in the perisylvian area containing Broca's and Wernicke's attested territories replete with neural tissue that as far as can be ascertained is the root of all human gab. Gab, however, is no guideline for the differentiation or qualification of individualism. Quite to the contrary.

We can't ourselves twittle beyond the game of thumb riot (two people, as if arm wrestling). In theory, outfitted with all the requisite equipment, someone white-gloved and steady of scalpel might be able to seize submicron-sized portions of their perisylvian and experiment with any and all forms of surgery, always aspiring to expand the region, retrofit its neural infrastructure, and thereby alter the allegedly preordained consortium of grunts, growls, cooing, and laughter, words not far behind. Students of psycholinguistics know every major paradox of language, from the outside. We have no virtual threshold by which to embark upon a fantastic voyage that would guide us in real time, obfuscating even the first base. A homerun would require some omnipresent vantage for assessing that moment during which time we could do away entirely with language. Many have embarked upon a period of silence and, with an equally zealous penchant, attempted to forget any and all words, subdue the impulse to speak, wrestle day and night with the temptation to cease all linguistically beholden thought. The variety of vows of silence give us fair warning that there is at large from time to time a sincere craving to expunge those habits of mind which have come down to mundane words and word construction. Go for one week without saying a word, and the susurrations seem to rejoice in the absence of a vastly dominant archetype.

We (the Authors) have attempted more than a few times to deliberately erase all language from our minds. It is not complicated to embrace silence, but astonishingly difficult to turn all of one's attention away from relations whose chatter reminds us persistently of a Weltanschaaung wherein every catechism hurls verbal circumlocutions at our monastic moments under the Sun.

The personages and "living agents" we have been hinting at are evolutionary magnets for any number of relationships, be they co-symbiotic or ecologically otherwise. The combinatorial power of those differing connections is fascinating and not a little bewildering. They are also crucial to the extent they are allied in various wise, though this cannot be established with certainty given the variety of taxonomic definitions, concepts, and memes that are hounded, distracted, and largely embedded in the various languages – arguably 7100 human languages alone spoken as of 2017.[23] In other words, the scientific disciplines that endeavor to understand the vastness of biological relationships are, themselves, hampered by the inhering bias of a language, whether English, German, Chinese, Russian, Japanese, or some translation modality, that cannot grasp any architectural concept of thought past its own minute sphere of confirmation biases.

Mindsets empower paradigms not easily altered. Our goal, in setting forth some of the rubrics that constrain our ability to gaze into a biological crystal ball, or transcend traditional strictures that have served to enshrine the so-called doctrine of DNA as Lewontin subtitled his 1991 book, published by Penguin, *Biology as Ideology*[24] (a good case made against many of the genetically determinist claims of sociobiology), is to gauge some extent by which our alliances with other species are altogether crucial (1) to our future as humans, (2) our sanity, and (3) our survival. The biosemiosphere is the constellation of bright biological stars that gives us to wonder why on Earth we have not embraced all the Others – possibly 100 million + other species – with imaginative and gentle enthusiasm since time began. And, were we to commence doing so, what the world might look like.[25]

In postulating that our minds – irrespective of the contentious ToM (Theories of Mind) debates – are no more precocious, insightful, or comprehending than those of any other Beings, we have a duty as ecological citizens of a living Earth to, at the very least, do our share.

What does that mean? Whether one embraces "amidst" or "among" (connoting amid our same kind or among others of other kinds), the quantities, multitudes, volumes and disparate natures of krill, microorganisms, the diversity of orchids in Borneo, lichen in the Antarctic, or the number of trees and shrubs on the planet on any given second are all equally referential and crucial to the moral basis for ecological duty. A good example derives from the current literature on the importance of perpetuating vast numbers of bacteria in our gut – good bacteria propagated by various pre- and probiotics – to help avoid the onset of Alzheimer's disease by encouraging with diet and exercise the cleansing of those "tangles" associated with amyloid plaque in the brain.

But with this phrase comes not platitudes and simply the self-serving survival regimens but real-time opportunities for pleasure, for feeling the sublime, for engaging in nonviolence. Compassion and the embrace of mutual interdependencies leading to a global course of biological equanimity are first and foremost in this mission to stem the tide of the escalating Anthropocene. That constitutes a ToH – a Theory of Heart surpassing intellect, with the ultimate goal worthy of this or any human generation.

---

[23] See https://www.ethnologue.com/. Accessed 9 Mar 2017

[24] For a recent review, see https://www.independentsciencenews.org/health/biology-as-ideology-the-doctrine-of-dna/. Accessed 9 Mar 2017

[25] See Tobias MC, Morrison JG (2017) "Epiphanies of the Biosemiosphere" Chapter 9. In: Anthrozoology: Embracing Co-Existence in the Anthropocene, Springer International Publishing, Cham, Switzerland, pp 225–332

ToM versus ToH (not that they cannot be precisely in conformity) might explain disease etiology and teleology in humans, the health of Charles Darwin being a case in point. It has been suggested that "Darwin's later health problems were caused, in part, by claustrophobic exposure to arsenic and formaldehyde as he worked on his specimens in the cramped *Beagle* quarters…".[26]

New Mind, New Infrazoology

Prosociality, that inclination – learned or otherwise adopted for any number of conscious or preconscious reasons – toward an engagement in the well-being of others, or of the entire community in which one lives or is focused upon, has been extensively studied, in the form of general altruism among humans. It was even mapped in the city of Binghamton, New York, by David Sloan Wilson and colleagues. Writes Wilson, "Most people are behaviorally flexible and can calibrate their prosociality to their circumstances," which is, of course, a most well-boding insight, if it is, in fact, the case.[27] Wilson also points out, in a footnote enriching this cited statement, that "Physical traits can be flexible in the same way. The general term that evolutionists use for traits that vary in response to the environment during the lifetime of the organism is *phenotypic plasticity*." And Wilson adds, "A large literature exists on phenotypic plasticity that is highly relevant to understanding and improving the human condition."[28] As another scientist named Wilson points out (Edward O. Wilson), "Examples of favorable mutant genes are those that prescribe adult lactose tolerance."[29] Conversely, "an unfavorable mutant gene in humans is that which prescribes cystic fibrosis."[30] Compellingly, in his important Appendix, E. O. Wilson takes up the controversies surrounding "inclusive fitness theory" in which he argues that the conventionally invoked "regression method" (that mathematical model for understanding and statistically modeling relationships amid variables – the seemingly perfect approach to all quantitative data whirling around evolution in general) in fact, in Wilson's opinion, "fails to distinguish between correlation and causation," that it is actually "useless for the prediction or interpretation of evolutionary processes."[31] And because, writes Wilson, the "Regression Method Does Not Yield Predictions," then "it is logically impossible to predict the outcome of a process without making prior assumptions about its

---

[26] See Haupt LL (2006) Pilgrim on the Great Bird Continent – The importance of everything and other lessons from Darwin's Lost Notebooks. Little Brown & Company, New York, p 199

[27] See Wilson DS (2015) Does altruism exist? culture, genes, and the welfare of others. Yale University Press/Templeton Press, New Haven/London, p 118

[28] Ibid., p 157

[29] See The Meaning of Human Existence, Liveright Publishing Corporation/A Division of W. W. Norton & Company, New York/London, 2014, p 62

[30] Ibid., p 62

[31] Ibid., p 190

behavior. In the absence of any modeling assumptions, all that can be done is to rewrite the given data in a different form."[32]

That would also invalidate the sanguine lessons allegedly divined from the pro-social map of Binghamton. Indeed, the very nature of understanding and assuming altruism, of goodness in humankind, becomes a contested variable within a complex of evolutionary uncertainties, with respect to our being able to understand them.

This is all the more likely when we consider the earlier referenced dominant memes of human cultures at great variance. Consider what happened when "the Mona Lisa" was finally allowed to leave the Louvre, and over 1.6 million visitors came to the National Gallery in Washington and Metropolitan Museum in Manhattan during January of 1963 to gaze fleetingly. According to Donald Sassoon, *"The New Yorker* [magazine] calculated that each of the 1.6 million visitors took an average of four seconds to contemplate the portrait." If this seems a likely precursor of the age of the Tweet, it assuredly is. Nonetheless, went the story, "That timing might seem a little rushed, but, as with pilgrims viewing a holy relic, it was sufficient to make people feel sanctified by the experience."[33] Imagine we humans so vulnerable as that to an iconic experience, which is certainly one plausible scenario. Think of all those many instances when we, too, have joined the rushing throngs at various major museum exhibition splashes, or wandered The Great Wall, or walked across Saint Mark's Square in Venice on a Summer's afternoon amid the happy hour of massive crowds. Or at Coney Island. We've all certainly experienced the sudden rush of some version of euphoria. It is instantaneous, like the proverbial 40 s John Ruskin ascribed to our maximum time of pleasure in smelling an orange; the same 40 s often attributed to the normal duration of a human orgasm.

A similar phenomenon occurred between November 1976 and April 1979, when the "blockbuster" museum tour across America of "Treasures of Tutankhamun" seduced many millions of visitors.[34] A few seconds of interest. Yet other studies have shown (in the case of environmental education deriving from children learning about endangered species in captive situations, e.g., zoos) that there are only a few minutes at most, out of many hours of the day during which anything remotely educational or inspiring sinks in.

Human nature is thoroughly discursive, meandering, rambling. But not necessarily random. And this fact butts up against any evolutionary theory of strict ordinance or adherence. Were we so chatty 30,000 years ago? Yes. By all evidence from the painted word: run-over by gossip.

Over 45 years ago, I (Michael) was exploring a wild gorge an hour from Neufchâtel. I climbed up to a vast thicket beneath a cliff, and bush-wacked to the

---

[32] Ibid., p 195

[33] Sassoon D (2006) Leonardo and the Mona Lisa Story – The History of a Painting Told in Pictures. An Overlook Duckworth/Madison Press Book, New York, pp 267–268

[34] See "King Tut: A Classic Blockbuster Museum Exhibition That Began as a Diplomatic Gesture," by Meredith Hindley, Humanities, September/October 2015, Volume 36, Number 5, Humanities – National Endowment For the Humanities, https://www.neh.gov/humanities/2015/septemberoctober/feature/king-tut-classic-blockbuster-museum-exhibition-began-diplom. Accessed 9 Mar 2017

rock face, wherein a cave invited me in. I took off my tee shirt and lit it afire, wrapped around a stick. For approximately 30 s, traipsing inside the cave, it was within seconds that I first glimpsed the cave paintings all around me. Then the light went out. I repaired to the sunlight and never returned, never to mention the event again. To my knowledge, to this day, no one has ever "rediscovered" this cave I have sought by my silence to protect.

What were those paintings? Well, I had seen a bison, or some such bovine. Was there also a wolf? A deer? Birds? A leopard? I cannot say for certain. It was so fast, so amazing. A dream I have kept to myself for two/thirds of my life. A bit like a lost Mondrian.

But in those days I was also constantly in touch with the papers and books of the great André Leroi-Gourhan whose 1967 publication of *Treasures of Prehistoric Art*[35] is so monumental, timeless, that the publisher chose not to date it. There has never been a more seminal work on Paleolithic aesthetics. For many of us, Leroi-Gourhan was the ultimate window on the workings of an enigmatic human history. His thinking and hand- and footprints were impossible to follow. My own accidental encounter with an as yet "unknown" cave (how silly an egotism) gave me some 30-s sense of the exhilaration and responsibility that came with the encounter of primordial human artistic sensibilities.

Between 1834, with A. Brouillet's discovery of a painted bone in the Le Chaffaud cave from the region of Vienne in France, to the discovery in 1940 of Lascaux, scores of what were then called "prehistoric" petroglyphs and other relics from across the world gave far greater pause than 4 or even 30 s to contemplate meaning, subtext, the geography, and potential cultural implications of a proliferation of zooglyphs. Whole cosmologies emerged from the Franco-Cantabrian Paleolithic surge of artistic expression, along with equally enigmatic but emphatic cave painting subjects from Australia, Amerindians, and across the Central Steppes of Asiatic Russia, as well as from Japan. The many animal figures, along with crystals, fossils, skulls and middens, flints, tools, beads, tiaras, and an ever-growing agglomeration of other remnants of human life in concert with that of dozens of other species, some now extinct, made for an entirely new prescience, of philosophy driving science.

"Nothing obliges us to suppose that Paleolithic men were different than us," wrote Leroi-Gourhan.[36] What remains from the summaries in the late 1960s of the thousands of images, objects, and sites that had thus far been discovered were questions, not answers, arbitrary styles, unclear dating, and motives that appear to have two universal forms: "symbolic and realistic."[37]

By the time of publication of Leroi-Gourhan's masterwork, he had personally viewed "2,188 animal figures distributed in the 60 decorated caves or rock shelters…. Among 110 sites."[38] He inventoried the animals, coming up with "610 horses,

---

[35] Harry N. Abrams, Inc., Publishers, New York
[36] Ibid., p 44
[37] Ibid., p 103
[38] Ibid., p 111

510 bison, 205 mammoths, 176 ibexes, 137 oxen, 135 hinds [females of the red deer], 112 stags, 84 reindeer, 36 bears, 29 lions, and 16 rhinoceroses."[39] And there were in addition other horned deer, "3 undefined carnivores, 2 boars, 2 probably chamois, and 1 probably saiga antelope."[40] There were also birds, fishes, and 9 monsters, including a unicorn at Lascaux, and possible giraffes. In addition, scores of humans, men and women, as well as ghosts.

Such accounting across numerous codified styles and chronological frameworks from the Urals to Spain, at a time when, according to Leroi-Gourhan, "around 20,000 B.C., the region extending from Russia to the Pyrenees seems to have constituted a vast cultural whole, with Czechoslovakia and Austria providing the transition between east and west" (as indicated by the carbon-14 measuring technology applied in the late 1960s); and also accounting for a post-glaciation period of moderation sweeping the entire geography in question, there still remains a fundamental question.[41]

Despite a uniformly awesome artistic shepherding of varied acuities and sensibilities, of true physiolatry, biophilia, sympatric aesthetics of astonishingly complex and detailed artwork displayed for reasons and in any number of canyons, grottos, rock faces, and within dark labyrinths of stone whose purpose will never be fully grasped, there remain innumerable questions at the heart of philosophical discourse. For example, in assessing hundreds of rock art sites throughout approximately 150,000 square miles just in the state of California, paleontologist Campbell Grant has stated, "Ethnological studies have shown that identical motifs often have completely different meanings in widely separated regions."[42] And as Robert F. Heizer and C. W. Clewlow, Jr., have pointed out in their own extensive two-volume analysis of California prehistoric art, with respect to categorizing a rich repository of images, they write, "'Human' and 'Animal,' deal with figures, and refer to any human or animal figure, or portion thereof... Quite possibly some very stylized elements intended by their original makers to portray insects or supernatural creatures were not recognized by ourselves as such. If this be the case, they are hidden in other categories, and we can only hope that more sensitive observers than we are may someday rescue them from their incorrect assignment."[43]

Of particular and equally baffling interest is the very naming of animals, totemic spirits, and thus the frequency and likelihood by which posterity is inclined to categorize and think of our predecessors. A case in point comes from the very "origin of names" as described in Frederica de Laguna's three-volume work on the Yakutat Tlingit culture, one of the great masterpieces of global ethnographic research. Laguna described how "personal names" might derive from animal emblems that

---

[39] Ibid

[40] Ibid

[41] Ibid., p 204

[42] Campbell Grant, "Rock Art in California," *The California Indians*, edited by R. F. Heizer and N. A. Whipple, p 233

[43] Heizer RF, Clewlow Jr. CW (1973) Prehistoric Rock Art of California, vol 1, Ballena Press, Ramona, California, p 9

had been adopted – or adapted – by a clan; from ravens and small ducks; or, in the case of grandchildren named at potlaches, like the Thunderbird. But that was a mere prelude to a veritable infrazoology of familiar local species and elements of nature including the "(Eagle Crying Around a Dead Man…Dust (woman), Picking Strawberries (woman), "Swimming Wolf," (Bear) Showing his Face Outside, (Bear) Entering his Den,"Killerwhales Tearing up the Water, [or] Continually Jumping on One Another…No Eyes, Is Never Lost, Clean(?) Around the Mouth, Alongside Itself (woman), Heavy Wings, and Sealion Tooth…Old Frog…Salmon…Tall-Tree Father…Dried-Hand Father, Easily-Blown-Away Father…Frog's Daughter…pretty woman…" and so forth.[44]

Even in cases of relative modernity, such as the enormously detailed Cheyenne watercolor on paper with colorful inks and graphite, entitled "Sun Dance encampment," possibly by "Little Chief" (59.5 × 65.8 cm), the hundreds of meticulously choreographed figures retain the same mythic and ritual anonymity as those tens of thousands of players in the drama engendered by Albrecht Altdörfer's "The Battle of Alexander at Issus" (1529) Wood, 158.4 × 1120.3 cm, at the Alte Pinakothek in Munich. While the Cheyenne piece "illustrates the many social and sacred activities associated with the Sun Dance," we are left with an overabundance of questions about identity: Who were these beautifully intimated forces of life, embodied in the guise of so many solitaires sharing the canopy of a social existence?

Unlike the many hundreds of historically specific individuals photographed by Edward Curtis, the overwhelming lot of that artistic and emotional expression of humans and other animals leaves no trace in terms of names, genealogy, or family history.[45] This anonymous through-story is indeed the story of billions of years of biology. Even "several animals in [Hieronymus] Bosch's painting ["The Garden of Earthly Delights," ca.1500] "one of the most enigmatic and interpretively contested paintings in the history of western art" are wonderfully bereft of "any classification, and result therefore in 'empty' or aberrational memories'." This, on account "of their fabulous forms."[46]

When we examine the vast panoply of brilliant styles of aesthetic and ecological release and specific paintings and sculptures from throughout history – with the exception of works largely confined to the early Middle Ages and subsequent origins of the Western Renaissance and all those signed pieces subsequent to it – a similar namelessness is almost predictably inherent to Egyptian or Mesoamerican pyramids and Central Asian thangkas, alike. "Japan stands out in Asia as an exception," wrote

---

[44] *Under Mount Saint Elias: The History and Culture of the Yakutat Tlingit*, by Frederica de Laguna, Smithsonian Contributions to Anthropology, Volume 7, In Three Parts, Smithsonian Institution Press, City of Washington, 1972, Part Two, pp 788–789

[45] See *Native American Painting – Selections from the Museum of the American Indian*, by David M. Fawcett and Lee A. Callander, Photography by Carmelo Guadagno, The Museum of the American Indian, New York, NY, 1982, p 24

[46] See Falkenburg R (2011) *The Land of Unlikeness: Hieronymus Bosch, The Garden of Earthly Delights*. WBooks, Zwolle, The Netherlands, pp 7 and 122

the famed art historian Jacques Élie Faure (1873–1937) in his *The Spirit of the Forms*, part of his glorious *History of Art* series.[47]

If we consider "the world's most famous illuminated manuscripts" from AD 400 to 1600, the title of Ingo F. Walther's and Norbert Wolf's book by that title[48] makes it abundantly clear that all of the many psalters, bibles, rolls, coronation gospels, evangeliars, a profusion of codeces, commentaries, Books of Hours, Responses, Apocalypses, historiariums, missals and sacramentaries, Narratives, letters, initials, journeys, Adventures, calendars, collections and enchiridions, collected poems and breviaries, folios and parchments, rosaria, moralia, chronicles, Miscellanies, stories and books, genealogies, exploits, manuscripts and sacred texts, scriptoria, periscopes and lectionaries, single leafs and products "from the hand of..." are teeming with names: authorship, subjects, translators, writers, those profiled, those in power, those who have commissioned the works, those to whom the works are dedicated, saints. Real identities. People in chronicled history. This competes with Adam's naming of the animals. These are real personages who have entered history on the printed, illuminated pages of ethics, ideals, and commiseration. In no other genres, eastern and western, have we as a world cultural meme inherited so much beauty and precision, marking a major transition from anonymity to authorship or provenance. This is one of the least discussed and most glaring psychoanalytic transitions in recent human evolution: the personification of the ego.

To cite but one utterly astonishing high point in the miniature art form so designated: the so-called Rohan Master, his Grandes Heures de Rohan, ca. 1430–1435, Paris or Angers, 239 folios in Latin, in the Bibliothèque nationale de France, Ms. Lat. 947.[49] Amid howling dogs, anxious sheep and razor sharp limestone cliffs is folio 135 in which Christ lies on the ground decomposing, grief is everywhere, and even God is painted in a state of unrequited despair, his identity fully comprehensible. Write Walther and Wolf, "No other passion scene in the history of mediaeval book illumination has anything approaching the high drama of this unforgettable miniature."[50] But what we also must note is the aggrandizement of the personages. The artist(s) have ensured the perpetuity of real individuals. One's mind goes immediately to a Pope Francis, to a Saint Francis, or to Father Mychal Judge, the Chaplain of the New York Fire Department who was among the first to perish at the World Trade Center on 9/11.

What a contrast with the only partial individuals, all those tormented and in various stages of fragmentation and deconstruction painted by a Francis Bacon, as most disturbingly revealed in a Portuguese exhibition of works in 2003.[51] In the interview with David Sylvester published in the catalogue, Bacon speaks of a

---

[47] Translated from the French by Walter Pach, De Luxe Edition, Garden City Publishing Co, Inc., New York, 1937, p 74

[48] Taschen, Köln, 2005

[49] See Walther IF, Wolf N (2005) Masterpieces of Illumination – The World's Most Beautiful Illuminated Manuscripts from 400 to 1600. Taschen, Köln, pp 307–309

[50] Ibid., p 309

[51] *Caged. Uncaged.* January 24–April 20, 2003, Fundação de Serralves, Porto

Rembrandt self-portrait: "…as it happened in this Rembrandt self-portrait.. there is a coagulation of non-representational marks which have led to making up this very great image… abstract expressionism has all been done in Rembrandt's marks. But in Rembrandt it has been done with the added thing that it was an attempt to record a fact and to me therefore must be much more exciting and much more profound… You see, I believe that art is recording; I think it's reporting…"[52] Reporting in no less a manner than those remarkable illuminations of Dante's *Divine Comedy* by Giovanni di Paolo.[53] Or Picasso's famous remark that when he is painting an object it is not to search for it but to find it. The personage has become ground zero.

The Individual Amid Multitudes

Edward Hicks rendered over 100 "Peaceable Kingdom" paintings. In each of which "…The wolf and the lamb shall feed together…"[54] These were among the earliest North American paradise primitives. But they differed little from the works of Giovanni Paolo, or any of the illustrations that garnish the history of that art we consider integral to civilization. But, as poet Robert Creeley intoned in his volume *Life & Death*[55] "The agonies of simple existence lifted me up. But the mirror I looked in now looks back."

In his massive overview, *A Perfect Moral Storm: The Ethical Tragedy of Climate Change*, by Stephen M. Gardiner,[56] the paradox of numbers translates with sublime forlorn into the modern ecological crisis as Gardiner contemplates "both ideal theory and the ethics of the transition,"[57] a transition that in its most paramount guise bears "witness to serious wrongs even when there is little hope of change."[58] And Gardiner continues, "We face a looming global environmental tragedy. Given that we see it coming, why has our response been so limited?"[59] We know which specific individuals, from country to country, political party to party, are guilty of denial or outright contravention. But there is no answer. Only that "The ethical challenge is unusually difficult, the stakes extremely high, and it will not be easy for us to emerge morally unscathed."[60] Our species, in other words, is held hostage by individuals.

---

[52] Ibid., p 235

[53] See Pope-Hennessy J (1993) Paradiso – The Illuminations to Dante's Divine Comedy by Giovanni di Paolo. Random House, New York

[54] See Hecht A (1995) *On the Laws of the Poetic Art*. In: The A. W. Mellon Lectures In The Fine Arts, Bollingen Series XXXV:41, Princeton University Press, Princeton, New Jersey

[55] A New Directions Book, New York, 1998, p. 68, from the poem, "Oh My God…"

[56] Oxford University Press, New York, 2011

[57] Ibid., p 437

[58] Ibid

[59] Ibid., p 439

[60] Ibid., p 442

The transition to an activist stance in the human history of art has been no less arduous. But at some point, we moved from that anonymous Everyman to individuals willing to sacrifice themselves for a cause célèbre. The first signature in human history – the Sumerian scribe, one Gar Ama from approximately 3100 BC?[61] The first signed garden – Ryoan-ji – in Kyoto, late fifteenth century, by Sôami? The first signed contract? Work of art? Autograph? Signed letter? Tracings in the sand? Petroglyph?

If we feel that somehow incredibly we have in just the last few years transcended this impasse – from passive dot in the thicket to a forceful voice with a name, a mission, and a destiny – think back to the American election in early November 2016 or to the Moravian brethren and Missions in the nineteenth century Greenland, a concatenation of persons who have vanished amid the melting ice, beneath thawing gravestones and the wandering terror in their hearts of polar bears.[62]

Danish and other families will remember these graves, and those who struggled there and now inhabit that land. They are not anonymous, not in the way that William Blake famously envisioned the life of a fly, "If thought is life/And strength & breath,/And the want Of thought is death;/Then am I/A happy fly,/If I live/Or if I die."[63]

But by comparison with American politics in 2016/2017, the early settlers of Greenland comprise a relatively calm and easily digested scope of human affairs, with questions and actual answers; identities and real lives. Whereas the democratic election of dictators who have the capacity to destroy the planet with unprecedented technology represents a grievance unknown in the annals of at least 64 millions of years prior to now and would appear to give rise to an individual destined to kill Others.

This situation, resoundingly, calls upon the individual within the species, whether we have philosophically or scientifically rallied to the point of ascribing individualism or not. Our dilemma is that we must and that *we* have summoned an astonishing arithmetic composed of single numbers, one by one, personage upon personage, to act out upon the stage of life her/his truth; a truth of being that will not be stayed, is committed to some kind of motivated happy-ending change and, by all the actions and actionable causes, has the preparatory bearing and communitarian convictions of an organism destined to see reality transfixed, nature appropriately protected, the Other voluntarily engaged. Our meaningful and courageous communion at this point in time represents the crucial moment in our history.

---

[61] "The History of Signatures," by Legalesign Staff Writer, 19 February 2016, https://legalesign.com/blog/history-of-signatures/. Accessed 10 June 2017

[62] See *Cultural Encounters at Cape Farewell – The East Greenlandic Immigrants and the German Moravian Mission in the nineteenth century*, by Einar Lund Jensen, Kristine Raahauge and Hans Christian GullØv, Museum Tusculanum Press, University of Copenhagen, 2011

[63] *A Bestiary*, Compiled by Richard Wilbur, Illustrated By Alexander Calder, Pantheon Books, New York, "The Fly," by William Blake, p 18

Chapter 7
Ecological Existentialism

The Poetry, Science, and Despair of the Anthropocene

A Danish physicist at the Niels Bohr Institute, Holger Bech Nielsen, has stated, "We do not even know if there should exist some extremely dangerous decay of say the proton which caused eradication of the earth, because if it happens we would no longer be there to observe it and if it does not happen there is nothing to observe."[1] This perspective on temporality is old news to philosophers and mathematicians. But the split-edge of time that defines, ignores, or eradicates all things human is worth recalling in view of the Anthropocene, provoking a very different kind of outcome, one more attuned to T. S. Eliot's poem, "The Hollow Men" (1925), with its much contemplated concluding stanza:

This is the way the world ends/This is the way the world ends/This is the way the world ends/Not with a bang but a whimper.[2]

Amazingly, in a later interview Eliot apparently questioned his commitment to this stanza by stating that he was "not sure the world will end with either [referring to war, or to nuclear holocaust]. People whose houses were bombed [during World War I] have told him they don't remember hearing anything,"[3] a veritable stamp of authenticity on the above-referenced remark by Nielsen. Similarly, several survivors of the catastrophes at Hiroshima and Nagasaki remarked upon the "beauty" of the light in the sky, as the atomic bombs burst, destroying nearly every living

---

[1] From: "Random dynamics and relations between the number of fermion generations and the fine structure constants" Acta Pysica Polonica B, May 1989), quoted from Canadian philosopher John A. Leslie's book, *The end of the world: the science and ethics of human extinction*. Routledge, London, UK, 1996, in Wikipedia, under the heading, "Human Extinction," Note #4 from a Wikipedia entry on "Human Extinction," Accessed 29 Mar 2017

[2] 23 November 1925, in Eliot's Poems: 1909–1925, Harcourt Brace, New York and Chicago

[3] "T. S. Eliot at Seventy, and an Interview with Eliot" in Saturday Review. Henry Hewes. 13 September 1958 in Grant p 705

creature around them.[4] How could Hitler's inner circle of monsters have been so public regarding their adoration of such musicians as Bach and Beethoven?

We have seen deaths of entire cultures, like those intimated from Tamaulipas, Northern Mexico, several thousand years ago. But, of an even earlier antiquity, and one continuing to this day, the Todas of Tamil Nadu, southern India, elaborately and elegantly defy the notion of human extinction, despite their diminutive population size (under 1500 individuals).[5] The Todas are a rarity. The most recent studies of massive die-offs of populations inherent to the Anthropocene are of such inordinately ponderous and vast interrelated numbers, exceeding by many powers of ten the alleged normal extinction background rate, as to suggest nothing less than a global biological crash, both physical and mental, from whose ruins few if any humans are likely to walk away.

There have been countless data sets from the tropics as well as the Arctic with respect to genetic repositories and their waning durability or robustness in the face of extreme climate disruptions.[6] In two very specific instances involving indigenous parrots of Tasmania (the swift and orange-bellied parrot),[7] their highly unusual migrations over water (the 310-mile-long Bass Strait separating Tasmania from the southern Australian state of Victoria) in addition to countless human and nonhuman threats have savaged both parrot groups' breeding numbers, particularly the latter. But even the swifts have been mathematically modeled to the extent that their extinction has been predicted by 2031.[8] This is rather astonishing news with respect to stochastics, the "oscillator momentum" data developed by George Lane in the 1950s as it pertained to Wall Street.

If we can predict a species' end, tying it to an actual human year, then we can do so for humans, as well. The difference between *Lathamus discolor* (the swifts) and humans is simply respective current numbers and the variables such numbers implicate.[9] In the case of the orange-bellied parrot (*Neophema chrysogaster*), a mere 50 breeding pairs remain. As with so many internationally critically endangered

---

[4] See "Voice of the Planet," the ten-hour television miniseries by M. C. Tobias and J. G. Morrison, Turner Broadcasting Service, Episode #8, "Extinction," 1991, based upon the novel by the same title by M. C. Tobias, Bantam Books, New York, 1990

[5] See Tarun Chhabra (2015) The toda landscape: explorations in cultural ecology, Harvard Oriental Series 79, Harvard University Press, Cambridge, MA

[6] For example, see Bellard C, Bertelsmeier C, Leadley P, Thuiller W, and Courchamp F (2012) Impacts of climate change on the future of biodiversity. Ecol Lett 15(4): 365–377. doi: https://doi.org/10.1111/j.1461-0248.2011.01736.x, PMCID: PMC3880584, EMSID: EMS54918, https://www.ncbi.nlm.nih.gov/pmc/articles/PMC3880584/ Accessed 30 Mar 2017

[7] See https://www.difficultbirds.com/orangebellied-parrot, Accessed 30 Mar 2017

[8] See "Tasmania's swift parrot in danger of extinction, calls to list the bird as critically endangered," 29 May 2015, http://www.abc.net.au/news/2015-03-26/research-shows-swift-parrot-in-danger-of-extinction/6350564 Accessed 30 Mar 2017

[9] See Heinsohn R, Webb M, Lacy R, Terauds A, Alderman R, Stojanovic D (2015) A severe predator-induced population decline predicted for endangered, migratory swift parrots (Lathamus discolor). Biol Conserv 186: 75–82. Available online 25 Mar 2015. https://doi.org/10.1016/j.biocon.2015.03.006. Accessed 30 Mar 2017

species – from pupfish (*Cyprinodon diabolis*) in southwestern Nevada's Ash Meadows National Wildlife Refuge to the southern muriqui (*Brachyteles arachnoides*) in Brazil, a single localized incident could easily drive them to extinction or zombie-like extinction (the walking dead).[10]

Once the conversation begins, there's no stopping it.[11] Are we so different from two parrot species in Tasmania? They, like at least half the world's biodiversity – its infinitude of remarkable individuals – are being swept away, anonymous castoffs in the human-induced whorls of a holocaust that knows no end until, indeed, it is over.[12] Countless researchers backed by typically corporate underwriting have sought to manipulate the approximately 20,500 human genes, each gene locus having been identified, in an effort to accomplish concerted manipulations of individuals, a financial repudiation of extinctions. But, again, a chimera. By April 2003 the International Human Genome Sequencing Consortium ("HGP") had published the first complete human genome. Other species – from fruit flies to sheep, from flatworms and roundworms to mice – have also been entirely sequenced. To what end? The answer does not exactly roll off the lips. The myriad categorizations of the universal principles behind "a model organism" have enabled such newly applied techniques as a variety of homologous gene therapies, cloning, DNA sequencing, artificial chromosomes, and polymerase chain reactions. But in the end, poets will write of stone.

The latter-referenced technique, PCR, provides within a matter of hours for the amplification of DNA configurations, up to tens of millions of target DNA sequences per minute.[13] In other words, the means of *manufacturing*, basically, new individuals. But they are not individuals. Rather, the products of a manufacturing process.

---

[10] See sequences in the Mata Atlantica [Brazil's Atlantic rainforests] in the movie, "Hotspots," produced by M. C. Tobias and J. G. Morrison, PBS, 2008. See also Dr. Paul R. Ehrlich's talk "The present mass extinction: how do the Tropics fit?" at James Cook University | The division of tropical environments & societies: centre for tropical environmental & sustainability science, 23 Nov 2016, published in MAHB – Millennial Alliance for Humanity and the Biosphere, http://mahb.stanford.edu/library-item/present-mass-extinction-tropics/. Accessed 30 Mar 2017

[11] Drake N (2015) Will humans survive the sixth great extinction?. Natl Geogr http://news.nationalgeographic.com/2015/06/150623-sixth-extinction-kolbert-animals-conservation-science-world/. Accessed 31 Mar 2017. See also Auerbach D (2015) A child born today may live to see humanity's end, unless…. Reuters, http://blogs.reuters.com/great-debate/2015/06/18/a-child-born-today-may-live-to-see-humanitys-end-unless/. Accessed 31 Mar 2017. See also, Proc Biol Sci. 7 Sep 2006; 273(1598): 2127–2133. Published online 2006 Jun 8. doi: https://doi.org/10.1098/rspb.2006.3551, Journal List, Proc Biol Sci, 273(1598); 2006 Sep 7 PMC1635517, The Royal Society Publishing, Proceedings B, https://www.ncbi.nlm.nih.gov/pmc/articles/PMC1635517/ Accessed 31 Mar 2017, PMCID: PMC1635517, Davies RG, Orme CDL, Olson V, Thomas GH, Ross SG, Ding TS, Rasmussen PC, Stattersfield AJ, Bennett PM, Blackburn TM, Owens IPF, and Gaston KJ Human impacts and the global distribution of extinction risk

[12] See "Half of Global Wildlife Lost, says new WWF Report," https://www.worldwildlife.org/press-releases/half-of-global-wildlife-lost-says-new-wwf-report, World Wildlife Fund issues 10th edition of "The Living Planet Report," a science-based assessment of the planet's health, September 30, 2014. Accessed 31 Mar 2017

[13] See https://www.genome.gov/glossary/?id=159; see also NIH, https://www.genome.gov/12011238/an-overview-of-the-human-genome-project/

We should feel queasy about such prospects and the notion of creating so-called perfect individuals, whatever that would mean to the latest sci-fi epic. On January 23, 2017, scientists at The Scripps Research Institute/Madeline McCurry-Schmidt engendered "the first stable semisynthetic organism."[14] Naturally occurring genes have been rendered ineligible for patenting by the courts since 2013, but synthetic organism patenting is enjoying a Wild West acceptability.[15] With the recently employed SBML, systems biology markup language, as well as the SBGN, systems biology graphical notation, we can be assured that the manipulation of our genome and that of other species is just getting started, with absolutely no legal or ethical guidelines to ensure even the most elementary restraint or consideration.[16]

Synthetic or systems biology goes to the core of current applications and application thinking with respect to hybrid individuals, both the concept and surreality of human manipulation of *Homo sapiens* and other species (e.g., turkeys, fish, and panthers). The Florida panther, critically endangered and until recently thought to number fewer than 50 indigenous Floridian individuals, has had a jumpstart, of sorts (its numbers have allegedly doubled), but they are hybridized panthers (with Texas cougars), not endemic Florida panthers. We do not herewith claim to understand the philosophical difference or not entirely. How could we? How could anyone?

The APA, or American Poultry Association, recognizes "eight varieties of turkeys in its Standard of Perfection… They are Black, Bronze, Narragansett, White Holland, Slate, Bourbon Red, Beltsville Small White, and Royal Palm. The Livestock Conservancy also recognizes other naturally mating color varieties that have not been accepted into the APA Standard, such as the Jersey Buff, Midget White, and others. All of these varieties are Heritage Turkeys."[17] According to PETA, approximately 300 million turkeys are slaughtered each year, just in the United States, for food.[18] According to the USDA, the number is higher, 360.995 million.[19] So what are we to make of any "Standard of Perfection"? Perfect for slaughter? The best tasting turkey? A human being with the highest IQ? Or the most perfect, ageless skin?

Worldwide, the UN Food and Agriculture Department specifies the huge array of body parts likely to end up in any meal of turkey, or poultry in general, consumed by humans. It comprises "meat, offals, raw fats, fresh hides and skins"[20] and 691

---

[14] https://phys.org/news/2017-01-scientists-stable-semisynthetic.html. Accessed 2 Apr 2017

[15] See Bryn Nelson (2014) Synthetic biology: cultural divide. Nature 509:152–154. doi:https://doi.org/10.1038/509152a

[16] See Adam G Bower, Maria K McClintock, and Stephen S Fong (2010) Synthetic biology -a foundation for multi-scale molecular biology. Bioeng Bugs 1(5): 309–312. doi: https://doi.org/10.4161/bbug.1.5.12391, PMCID: PMC3037580, https://www.ncbi.nlm.nih.gov/pmc/articles/PMC3037580/. Accessed 1 Apr 2017

[17] "Definition of a Heritage Turkey," https://livestockconservancy.org/index.php/resources/internal/heritage-turkey. Accessed 2 Apr 2017

[18] http://www.peta.org/living/food/turkey-factory-farm-slaughter/. Accessed 2 Apr 2017

[19] http://usda.mannlib.cornell.edu/usda/current/PoulSlau/PoulSlau-03-24-2017.pdf. Accessed 2 Apr 2017

[20] www.fao.org/.../ess.../Livestock_statistics_concepts_definitions_classifications.doc. Accessed 2 Apr 2016

million "farmed" turkeys in total – but that was data going back to 2004.[21] This is a low-ball estimate. It comes nowhere close to the turkey's cousin, the chicken, whose species, *Gallus gallus* witnesses approximately 44.5 billion mortalities at human hands each year.[22] When a Washington Post reporter tried to investigate the number of separate individual cows and steers that were likely to end up in a hamburger patty, the number soon escalated to as many as 100, a far cry from the mere "16 hides" said to line the seats of the new $3 million Bugatti Chiron.[23]

Breeding, hybridizing, and environmentally altering the circumstances of birth, short life, and tortuous death have become the hallmark of global industrial animal agriculture. When we try to imagine an individual, the blur is a blasphemy of individual lives crammed cruelly into cages and on to assembly lines of murder. The vast majority of consumers never care to find out. *Care* is the operative word. People largely ignore the plight of those trillions (as counted in decades) that are martyred for throw-away dinner plates. Benjamin Franklin preferred the behavior of the indigenous North American turkey to that of the bald eagle, as deciphered from portions of a letter to his daughter.[24] Given the breeding frenzy engaged by humans against turkeys, it is unlikely that even had the turkey become America's national bird that people would have ceased to devour them, particularly over Thanksgiving and Christmas. So different, the Pilgrims from the vegetarian Essenes who occupied caves above the Dead Sea, between today's Israeli and Jordanian borders. But there it is. Eagles, on the other hand, have continued to enjoy various levels of conservation protection, like the Florida panther. But up in Dutch Harbor, Alaska, the largest fishing port in the United States (far out in the Aleutian Islands), much of the 1.8 billion pounds of pollock are processed each year – and that's just one species, the one that often ends up in sushi, known as "imitation crab."[25]

"Imitation crab" has become familiar to millions of consumers who by all indications appear to have no concerns over what they're eating. We've known vegetarians who consume this sushi believing that imitation somehow colludes with biological neutrality, neither implicating pollock specifically, or any fish, for that matter. But

---

[21] Statistics: global farmed animal slaughter, from farmed animal watch: Number 68, Volume 2. 9 Sept 2004. https://www.upc-online.org/slaughter/92704stats.htm. Accessed 14 May 2017

[22] "Intensive Farming and the Welfare of Farmed Animals," http://www.fao.org/fileadmin/user_upload/animalwelfare/intensive_farming_booklet.pdf. Accessed 2 Apr 2017

[23] "I tried to figure out how many cows are in a single hamburger. It was really hard – Pursuing the unsettling question of how many cows are actually in a hamburger." By Roberto A. Ferdman, August 5, 2015, https://www.washingtonpost.com/news/wonk/wp/2015/08/05/there-are-a-lot-more-cows-in-a-single-hamburger-than-you-realize/?utm_term=.ed67e59e7350. Accessed 2 Apr 2017. For information on the Bugatti, see The Los Angeles Times, www.latimes.com/business/.../la-fi-hy-bugatti-chiron-drive-20170609-htmlstory.html. Accessed 10 June 2017

[24] American Myths: Benjamin Franklin's Turkey and the Presidential Seal - How the New Yorker and the West Wing botched the history of the icon," by Jimmy Stamp, http://www.smithsonianmag.com/arts-culture/american-myths-benjamin-franklins-turkey-and-the-presidential-seal-6623414, Smithsonian.com, January 25, 2013. Accessed April 2, 2017

[25] See Nation's Restaurant News, "Alaska's most prolific fishery," Feb 21, 2013 http://www.nrn.com/blog/alaskas-most-prolific-fishery. Accessed 2 Apr 2017

the gorgeous *Gadus chalcogrammus*, a member of the cod family distinguished by their[26] diel vertical migration every 24 h, movement from sea depths to near surface waters, is considered the largest biological migration on Earth. Quite a superlative for a sentient being that is devoured unceasingly by humans who claim to be interested in the outdoors, in nature programs, in the wilderness experience, so-called. This famously ignored remarkable fish is for the most part destined for a bleak processing plant, a marine, ice engulfed gulag. I (Michael) thought of all those beautiful fish when I once met and had a long chat with Mrs. Solzhenitsyn (Natalya Reshetovskaya) in a Cavendish, Vermont grocery store, close-by where both she and her husband, and I, were living in our respective domiciles, all of us in a kind of exile. These pollocks are one piece of the not so puzzling nearly trillions of fish trapped in the human gulag of by-catch killed each year. In 2009 (said to be the most recent calendar year for which data was kept by NOAA) "Americans consumed a total of 4.8 billion pounds of seafood, or approximately 15.8 pounds of fish and shellfish per person (both wild-caught and farmed)...."[27]

The data as early as 2001 from Dutch Harbor, as well as the northeastern Pacific and eastern Bering Sea in general, from the North Pacific Groundfish Observer Program overview (in which "observers sampled 44,272 hauls for species composition, collected 48,992 age structures, took measurements of 1,112,045 fish and crab, checked 8,597 female crabs for the presence of eggs, and found eggs in 4,390 female crabs") implicates not merely the beautiful and benign pollock. Pacific cod and rockfish, Pacific halibut and other flatfish, Pacific whiting and sablefish, Atka mackerel and Pacific salmon, and crab and "other fish" were all part of the studies.[28]

Watching over this madness, at Dutch Harbor – the largest such port of slaughter in North America – are some 500 American bald eagles, one-tenth the population of resident humans there; the same avians Benjamin Franklin was unhappy to see outbid the wild turkey as America's national bird. We watched a flock of several hundred wild turkeys take flight just after sunrise in the Ozarks many years ago and remain utterly perplexed that this magnificent bird, the wild turkey, like Alaska's pollock, draws a veritable blank on the charismatic megafauna map that has supposedly been wired into America's consciousness. 500 or so bald eagles are tolerated, even celebrated at Dutch Harbor, while billions of fish are hooked, beaten up, and suffocated. No individual is recognized, except on a ledger sheet. The meals produced for humans are hybridized, just as the alleged increase in Florida panthers has depended in large measure – as a conservation tool – upon hybridization with those Texas cougars. When a mother and two kittens were filmed wandering just north of the Caloosahatchee River in Charlotte County, biologists were overjoyed, believing that the species had increased from a few dozen individuals to possibly more than 200.[29]

---

[26] http://funwithkrill.blogspot.com/2011/08/diel-vertical-migration-dvm.html. Accessed 2 Apr 2017

[27] See http://www.nmfs.noaa.gov/aquaculture/faqs/faq_aq_101.html. Accessed 2 Apr 2017. See also https://www.npfmc.org/. Accessed 15 May 2017

[28] https://www.afsc.noaa.gov/FMA/PDF_DOCS/NPGOP%20REPORT%20-%20Overview%20 2001%20-%20web.pdf. Accessed 3 Apr 2017

[29] See "Film captures good panther news, by David Felshler, Los Angeles Times, p A15, Sunday, 2 Apr 2017, p A 15

Hybridization has become a modus operandi form of the glib in conservation biological circles, but also in all things medical. For example, National Geographic reported in January 2017 upon a "Human-Pig Hybrid Created in the Lab…" and described researchers who had created "chimera embryos" in the goal of taking "steps toward life-saving lab-grown organs."[30] "In a remarkable — if likely controversial — feat, scientists announced today that they have created the first successful human-animal hybrids."[31] The accompanying photograph was captioned thus: "This pig embryo was injected with human cells early in its development and grew to be four weeks old."[32] The whole idea of chimeras, in this context, is tragic. Our friend Dr. Marc Bekoff in his column online for Psychology Today wrote of "monsters" in rallying to the protection of these human scientifically manipulated victims.[33] But, the monster is in the human, not the genes, nor collections of genes. Neither any "God" nor John Donne would conjure such demonic wagers and malicious intent from the prelapsarian, primordial muck. This is not Shakespeare's conventional Tom o'Bedlam, or "tomfoolery" unleashed in the laboratory. Add Mary Shelley's lightning to further enliven the transformation of the DNA of raw matter into the Übermensch and the picture is complete. An all-too-vivid mirror of one vast pernicious outcome of synthetic biological studies and engineering, these anonymes, as the ancient Greek language thought of such beings; their two French synonyms connoting "anonymous" and "insignificant."[34] There is no legal apparatus to offer the least protection to these hybrids. As Bekoff writes, "The Farm Security and Rural Investment Act of 2002 amended the definition of animal to specifically exclude birds, rats of the genus *Rattus*, and mice of the genus *Mus*, bred for use in research."[35] For the whole evolution of chimeras, avians, *Rattus* and *Mus* are only the beginning.

In an important essay, the late Gregory Kavka discussed the notion of "treating the creation of rational beings as a means" and the thought that "to treat human life as a commodity…constitutes misuse of the agents' power over the existence of future individuals."[36]

That is as much a biological concept as it is a political one. We do not seek to implore the issues surrounding human rights in the future (as unclear as asking to be struck by M. Shelley's lightning) nor to echo the "existence before essence" credo of Jean-Paul Sartre in his first novel *La Nausée*[37] and the general, obtuse, underlying

---

[30] By Erin Blakemore, January 26, 2017, http://news.nationalgeographic.com/2017/01/human-pig-hybrid-embryo-chimera-organs-health-science/. Accessed 3 Apr 2017

[31] Ibid

[32] Ibid

[33] https://www.psychologytoday.com/blog/animal-emotions/201608/the-emotional-lives-chimeras-challenges-anthrozoology, Published 12 Aug 2016. Accessed 3 Apr 2017

[34] http://www.linternaute.com/dictionnaire/fr/definition/anonyme/. Accessed 3 Apr 2017

[35] op. cit., Bekoff, Psychology Today

[36] Kavka GS (1982) The paradox of future Individuals, Philosophy & Public Affairs. 11(2):20, 111 on, http://faculty.smu.edu/jkazez/pap/kavka.pdf. Accessed 4 Apr 2017

[37] Editions Gallimard, Paris, 1938, first English translation, *The Diary of Antoine Roquentin*,

Being before an individual's being of Martin Heidegger.[38] Instead, we wish to address the imperative real-time philosophy of activism whose differentiation from logic, philosophy, and the ecological sciences merits a metaphysical hierarchy of priorities, if we are to sort out the likely options of human individualism within the evolutionary forces that perpetually rewrite the biological and hence mental prospects for every species. The Self-zealot, René Descartes' phrase, "je pense, donc je suis" ("I think, therefore I am"), from Part Four of his 1637 treatise, *Discourse on the Method*, was a dumbed down if canny zinger to enshrine the continuing, masterfully sinister vogue of self-interest, the dominant species whose arrogance has come in one-liners to define all of zoology and binomial nomenclature, as codified by Linnaeus.

For Nicolas Léonard Sadi Carnot, writing in 1824 of the maximum efficiency of heat engines, and thus enshrining the second law of thermodynamics – that entropy, the dissimulation of a field or fields of energy which in any isolated system escalates as a function of time – this energy-related insight also heralded the notion that humanity might find ever more efficient engines to thereby transcend universal laws of finitude, even go to Mars, hubris inflated. Fly like Icarus to the Sun, find medical cures to become immortal, and dominate the planet in such a way as the crisis has now been characterized, wherein human consumption actually requires five planet earths to sustain our levels of resource extraction, pollution, and exhaustion. Our exceptional ability, however tragic, to vastly exceed earth's carrying capacity is now our common currency, the mental paradigm of our age.[39] The superlatives and corresponding outrageous human behavior (so comprehensively out of sync with every other individual and species on the planet) have become throw-away expressions pertaining to nothing that matters, it would appear. Merely words mouthed like radio lyrics from an another age, when, in fact, these are symptoms of that dynamo of decaying civilizations that have transmogrified into a Group Mind that has no ethical center, no singular aspiration, not a shred of evidence to suggest that cruelty was rejected, injustice fought, or the human conscience considered worthy of emphasis, let alone practice. In such a cartography it makes little sense to hope for collective rationality. The ideals espoused by those millions of personages who did, in fact, comprise some facet of that otherwise unknowing species must somehow form a chorographic gradient, a vector-valued function of a map that defines the geography of souls – all those individuals who actually mattered – in a wild sense, to others of their own family, clan, tribe, community, and must always matter in terms of any portion of habitat; in any iota or scintilla of habitat: habitat in the sense

---

translated by Lloyd Alexander, John Lehmann Publishers, London, 1949

[38] *Sein und Zeit*, 1927, *Being and Time*, trans. by John Macquarrie and Edward Robinson, SCM Press, London 1992

[39] *Réflexions sur la puissance motrice du feu et sur les machines propres à développer atte puissance*, Bachelier Libraire, Paris, 1824. See also Roy Morrison's review of *Empire of things: how we became a world of consumers, from the Fifteenth Century to the Twenty-First*, Harper/Collins, New York, 2016 – "The empire of things: an ecological response," by Roy Morrison, http://www.ecocivilization.info/the-empire-of-things.html. See also https://www.amazon.com/ Sustainability- Sutra- Investigation-Roy-Morrison/dp/ 1590793870

of landscape, countryside, regionalism, biodynamics, and every ecological relationship and interdependency, but also in the manner of the habitat of mind, the domain of morality, either side of God, within whatever belief, or lack of belief, one might consider the prospect of an individual's worth to another.

One could easily trace the impact of the Swiss educational reformer, radically sensitive and pragmatic intellect, Johann Heinrich Pestalozzi (1746–1827) whose motto was "Learning by head, hand and heart"[40] upon the Swedish philosopher/sociologist Torsten Hägerstrand (1916–2004) to better grasp the physics and mathematics behind the geographer's approach to the individual-species differential equations that have burgeoned since the time of Linnaeus. Hägerstrand focused upon cultural migration and, most importantly, applied space/time/multidimensionality in mathematics to geography, engendering techniques for statistically understanding movement of individuals within larger groups – orientations which he summarized in tersely envisioned time-space cubes and prisms.

Underlying all of the transdisciplinary pyrotechnics (e.g., "time geography") has remained a single intellectual device for articulating a virtually impossible field, namely, the *ideal*, a concept that has come to connote the individual who harbors that ideal. In no other sphere than that of individualism does it make total sense. A culture cannot honestly be said to harbor an ideal for the very obvious reasons that mathematics, statistics, and quantitative analyses get rightly in the way. Think about the various morphologies herewith: One does not *analyze the individual*, but, rather, the group and its trends, histories, turnabouts, etc. One looks at, thinks about, speaks with the individual, not the group and certainly not the culture. Analysis, in psychoanalytic terms, is the gelid geometric encincturing of a non-relationship: The observer and observed, a relation dating, in its modern incarnation, to the Trobrianders and their disinterested observer, the anthropologist Bronislaw Malinowski (1884–1942).

Citing Ernst Mayr's classic work, *Principles of Systematic Zoology* (1969), David Ludwig suggests that "Mayr's biological species concept and its success in explaining the distribution of Malaria in Europe by distinguishing between two species in the Anopheles complex"[41] is but one seminal example of the importance, difficulty, and continuing relevancy of empiricism in its pursuit of what is termed the value-free ideal. This ultimately subjective aspiration is predicated upon a suite of uncertainties that cannot persuasively differentiate between those key elements in any ontology, namely, "pragmatism, positivism, realism [and] interpretivism."[42]

The importance of such differentiation to science has, in great measure, focused upon the ambiguous pertinacity of the "species" definition and its correlation to

---

[40] See von Raumer K *The life and system of Pestalozzi*, translated by J. Tilleard, Longmon, Brown, Green and Longmans, London 1855

[41] Ludwig D (2015) Ontological choices and the value-free ideal. SpringerLink, Amsterdam, p 10. https://philpapers.org/archive/LUDOCA.pdf. See also link.springer.com/article/10.1007/s10670-015-9793-3. Accessed 8 Apr 2017

[42] Dudovskiy J Research methodology. http://research-methodology.net/research-philosophy/ontology/. Accessed 3 Sept 2017

reality. Because all reality hinges upon perception, except as practiced by metaphysics – that premise underlying all philosophy which has taken up the perennial question of Being in and of itself – every Law of Physics, for example, is actually the outgrowth of a Law of Being.

Hence, the *First Law of Metaphysics* must entail that which most thoroughly commands verisimilitude at both the epistemic and (human) psychological level. Cliodynamic assessments of databases encompass a field that champions every conceivable event horizon through the amassing of comparable information from microsociology, archaeology, mathematics, physics, sociology, and numerous other transdisciplinary summations of human history. Two of its current leaders in this relatively new field are Andrey Korotayev[43] and Peter Turchin.[44] What makes the cliodynamic methodology necessarily a slippery slope are the unexpected insights and outcomes deriving from greatly varying cliometric anomalies generated (from no singular database or technique, but rather, provocative ensembles of information) so as to alter projections according to preordained outcomes (possibly researcher instincts).

Take but one example. In seeking to differentiate the individual from his/her/its species, we find that this is a particularly difficult task if the organism is asexual – accounting for at least 0.1% of all vertebrate species[45] and essentially all unicellular organisms, from the archaea and bacteria (tens of millions of species at a fantastic variety of variance) to a large portion of plants and fungi. In other words, the overwhelming majority of all life on earth, discounting the bio-computations so rigorously worked up for all those strictly *heritable traits* most commonly associated with the mechanism of natural selection, and the disconcerting truths inherent to our manipulative unnatural selections.[46]

We have also encountered the even more taxing question of the "Other Minds" problem. Both sets of uncertainty – gene/population differentiation and Other Minds quandaries – invite contemplation of what we would term *contributory collaboration*. This concept welcomes subsets of every conceivable and abstract variety, acknowledging that the long, evolutionary swathe from prokaryote to eukaryote,

---

[43] https://philpapers.org/archive/LUDOCA.pdf

[44] link.springer.com/article/10.1007/s10670-015-9793-3. Accessed 8 Apr 2017. See also op. cit., Dudovskiy J Research methodology and http://research-methodology.net/research-philosophy/ontology/. Accessed 8Apr 2017

[45] See his essay, Korotayev, et al A trap at the escape from the trap? demographic-structural factors of political instability in modern Africa and West Asia. Cliodynamics 2/2 (2011): 1–28, http://escholarship.org/uc/item/79t737gt#page-1, Accessed 8 Apr 2017. In addition, see *Complex population dynamics: a theoretical/empirical synthesis*, Princeton University Press, and Turching's 1998 book, *Quantitative analysis of movement: measuring and modeling population redistribution in animals and plants*, Sinauer Associates Inc., 2003

[46] Dawley RM (1989) An introduction to unisexual vertebrates. In: Dawley RM, Bogart JP (ed) Evolution and ecology of unisexual vertebrates, New York State Museum, New York, pp. 1–18, cited in. Roberto Barbuti, et al. (2012) Population dynamics with a mixed type of sexual and asexual reproduction in a fluctuating environment. BMC Evol Biol 12:49, doi: 10.1186/1471-2148-12-49, PMCID: PMC3353185, 8 Apr 2016

from brain to mind, from life and death to other lives and deaths, must surely animate a central pillar of realism, namely, the Being within the being. That is to say, a thorough juxtaposition of the general and the specific.

The *theoretical individual* has now found its voice, amid the billions of years of evolutionary ethers; half-truths apotheosized into the never-never lands of selves shorn of superego.

The question that remains – the one most central to what, above, we have called the First Law of Metaphysics – is to consider outstanding segments of and windows upon this deeply fragmented story line made up of personages along the entire way, nearly 4 billion years of life adventuring across every biome on Earth. For anyone who has lingered at cemeteries like Père Lachaise in Paris, the evocative truth of all those personages strikes at the heart of the poetics of life. That question is also embedded not in *who we are*, an insoluble biological conundrum, but what options for survival and dignity are available to our muddled collective and all of her parts. Hence, we arrive at a very comfortable, intimate, and familiar territory: the choices we make.

Chapter 8
Choices

Jerome's Choice

The question of the individual's sway over immediate environmental circumstances ultimately arrives at the foremost pillar defining one's existence: Is the alleged role of natural selection meted out upon the very *choices* effected by all those caught within the whorls of evolution? Is natural selection predicated upon the intuitive and mathematical edifice of decisions rationally, irrationally, impulsively or sluggishly, methodically or inspirationally honed, in the service of some reflex or idea pertaining to self-preservation? This has been the primary modus operandi of the engine of natural selection, as science and sociobiology have always understood it. Namely, reproducibility. But what if self-preservation is linked implicitly to something else entirely, far more important than *survival*, a word that cannot but evince the memory of *survivors* – a word connate with the holocaust, each one a vast and tragically missing individual. What if self-preservation has nothing to do with biology or ecosystems, but, rather, a cosmic instinct that glorifies only love, tolerance, and cooperation, whether within a Black Hole, at the core of dark background matter, in a dust mote, in every vertebrate and invertebrate, each of which can be said to have its own humanity. This word, humanity, is our linguistic deep referential, a permanent bridge between what we assume, during our best moments, about ourselves and our communities, and the biosynderetic (sic) connectivity which many people have grasped for millennia, first in the realms of biosemiotics and then across that whole landscape of conscience.

Synderesis is most frequently equated with St. Jerome's (347–420) commentary on the vision of Ezekiel's four animals and the Prophet's belief that Adam maintained his conscience, even having been relocated from Paradise. Ezekiel ben-Buzi, exiled to Babylon for the first two decades of the sixth century before Christ, is revered in Judeo-Christianity, as well as other numerous Near Eastern ethnic communities for his insistence on human *conscientia* and the four cherubs (winged angels, by inference, all biodiversity) that remained beside him. Those animals – eagle, ox, lion, and human, each bearing four wings and serving as the wheels that carried Ezekiel in a chariot; God having made him out to be as warrior on behalf of the nation of Israel (*Book of Ezekiel* 1:1–3:27) is at the heart of an historic divining rod that is pure schizophrenia. The warrior with a conscience. Is that the fitting metaphor of life on earth? For the humanity in every life form? Or a pathological double bind (as the late ecologist Gregory Bateson used the phrase) that offers no solace, no stable chemical state, nothing but turmoil?

Ecologists speak of ecological flux in recognition of that which is normal about a climax forest and how it achieves that maturity, about each and every eco-dynamic that is caught out in a whirlwind of boom and bust, overshoot, collapse, and regeneration. But if we are to venture near some semblance of prophecy in which our metaphors all gleaned from the living and nonliving worlds make sense; to divine a future being within that which is the true conscience, the humanity, the moral and requisite choices to be made in service of something greater than ourselves, how are we to imagine a new nature, an individual imbued both with pragmatic idealism and a praxis that avoids the ruinous biases of human evolution? We have posed this question in different guises (as Raphael and Rubens were so taken with Ezekiel and Leonardo, El Greco, Dürer, Bernardino Luini, and hundreds of other great painters smitten by St. Jerome and most insistently his friendly lion) repeatedly throughout this treatise, not by accident. St. Jerome, the patron saint of all literature, letters, translations, encyclopedias and by implication, all of human history that is kept for a purpose, is the individual that such history has focused upon. As is his lion companion, who had, according to the Golden Legend of Bethlehem, obtained a thorn in his paw which Jerome removed, the two of them forever after, best friends. After Jerome died at the monastery in Bethlehem, his choice to heal a lion manifested itself in the choice of the lion to protect the monastery's donkey. Loyalty. Faith. The choice of compassion. These are the most heralded and iconic traits ever so prolifically elaborated upon with respect to an individual in the history of Western tradition. The same could be argued of Christ, St. Francis of Mahavira in Jain tradition, and of Lao-Tzu among the T[D]aoists and humanists in general. In each case the human individual is the individual throughout the biosphere in every Being.

This Jerome, with his sprawling ascetic frame, long beard, wild hair, open book upon a rock beneath the customary crucifix (symbol of so many things to so many traditions), all immersed within the High Renaissance Giorgionesque encircling forest, watched over by the lion in his cave, was never more elegantly portrayed than in Lorenzo Lotto's painting, "The Patron Saint Jerome," (ca. 1520) that hangs in the Romanian National Museum in Bucharest. No pathology need be placated in such an image. It rings as clear and honest as a tuning fork.

Ontological Madness

No surprise that St. Jerome, as far as we know, left no offspring, only crossable bridges between species. For the majority of our kind, our supposed success at reproducing has been our near downfall. With a human population heading toward

12+ billion people, all those biologically reproductive individuals have sought consciously or not to tout their fertility rates in the name of ego and fleeting family heritage but against the ignored backdrop of the human-induced Anthropocene, as we have insistently striven to implicate throughout this lyric of contemplations (and footnotes).

Equally, if not more arresting is this: What if questions of fertility and the confluence of genetic quanta making for the conventionally noted barriers that define species (reproductive capacity or as with asexual organisms, some other set of surrogate biological qualities edifying a singular type of being) are *not* the true compass of evolutionary strategies? What if most fervent choices are free of natural selection? What if even our predilections (sex drive, superego, etc.), attitudes, all of those case-by-case moments that should accrue into a personality, an object of attention, craving, love or hatred, obsession or disinterest, ponderations and procrastinations, nausea, and poetry alike were capable of easily transcending the reflexes and involuntary nervous system-related precisions of a lifetime?

What if we are truly free? Or are unwillingly transfixed, condemned, as Jean-Paul Sartre so frequently argued? In Hazel Barnes' famed translation of Sartre, "Man is condemned to be free; because once thrown into the world, he is responsible for everything he does. It is up to you to give [life] a meaning."[1] It is a freedom of the worst kind. Without mercy or faith. Condemned to being ourselves without the aid of evolution, God, or of any forces other than our own mortality.

Finally, what if we truly are condemned to our own invention, our own selves, such that our livelihoods and mental trajectories exist bereft of any collaborative biochemical boost? Beyond Earth, in other words?

If we consider ourselves to be the cognitive or moral agents of individuals who stand apart from the great seas of that desperate struggle, in most cases, simply to survive, then what, statistically speaking, are we? Individuals, communities, collectives of communities? Do geopolitics engender boundaries more powerful than the sexual lines that are viewed by most people as lines that either may or may not be willingly crossed in the name of propagation? The same with speech, with legislation that roils from administration to administration, and across the erratic terrain of each generation – dicta of proliferation? Is altruism a choice, an impulse, a reaction, or something else entirely without known ontology or teleology despite all the empirical guesswork and observational bias?

If "choice" connotes free will (and it is not our intention to engage in or attempt to revivify a thousand years of scholastic debate on this topic), then it must abide by some other musculature than that implicated in every theology. It has to surpass all those barriers of custom and history that have been informed and shackled through proprietary community standards according to a grand labyrinth of imperatives working to constrain rather than liberate its progeny. The proverbial anchorite should, by this biological way of thinking, have no quarrel with the cenobitic community as he/she is not a part of it and therefore ideologically shoulders none of its

---

[1] *L'être et le néant* (*Being and Nothingness*), Gallimard, Paris, 1943, translated into English by Hazel Barnes Philosophical Library, New York, 1956

burdens. But if the monk, then, so too, the poet. And if a poet, then any artist whose recourse is lodged in his/her own soul. And if an artist, a scientist, a prime minister, a Fool. And so on ad infinitum, much like the touted nation of Switzerland, that "man in the Holocene" (Max Frisch's 1979 novella of introspection to the point of utter despair) which proffers gentians, cowbells, the false enchantment of Johanna Spyri's novel, *Heidi* (1881) and the largest per capita military stashes and standing guards (weapons hidden beneath haylofts and in bunkers under otherwise tranquil chalets) in the world.

In other words, following this dialectical catalyst of contraries in one, even the most liberated of individuals is caught in the cross hairs of ancestral and living ties. The life of a Mahavira (ca. 599–527 BCE), as described by Bhadrabahu (d. 298 BCE) in the famed *Kalpa Sutra*, is certainly no different.

This leads tortuously by argument into divisions between phylogeny and spirituality. Are souls connected? In some ethical traditions, like that of the Jains, souls, Jivas, are part of a vast complex, each one free and individual but also interdependent in the ecological, chemical, and geological sense. When we start speaking of a life form in the guise of so many grains of sand or stars in the universe, we lose our thrust because infinities do not have the emotional standing of individuals. A half-million victims of a genocide will typically receive less attention than a few victims. We identify. This is the heart of theater, but also the reality that a person falls in love with another person, not a crowd. And that may be the only viable and vivid true compass reading we need to rally around: identification with another or, in another far more timely and relevant sense, empathy with others. But of course, we all know that.

Falling in love may be less interesting, scientifically, than never falling in love. Marriage is less compelling than divorce. The choice to reject human society by some mental stance or physical distance may well prove that choices have a neural or psychological infrastructure that is, by definition, free. Accordingly, such a proposition would have no end point, no conclusion, no proof, in an environment whereby natural selection remains the only prescribed orientation for those mutations, extinctions, and the survival of future offspring. Anxiety, terror, horror – these by all their disastrous turns lead nowhere in terms of population genetics. Here before us is lodged a disconcerting lemma or ambivalent postulation that propounds a very axiom of freedom absent all biological constraints. Again, in the manner of Sartre, we are free, whether one thinks of such freedom as a condemning finality or some new lease on life. But we still have no consistent ideas as to what freedom means. Can it be positive?

All is a gambit galvanizing our daily breath. A degree of horror mollified by life herself, which is what we know and everything upon which we can possibly speculate, and from which seek our only true and lasting satisfaction and joy. The oldest cliché: The brightest stars in the darkest nights.

Such speculation has entailed a history of histories, and we have forgotten every name, every definition. We are, in that sense, free, free to be alone, together, lost or found, doomed or saved. The drama queens of our peregrinations are combined, a question that remains forever and impossibly unanswered. Who, what are we? The query provokes endless riddles colored by Gauguin and steeped in both Socrates and Shakespeare, enlivened by such perplexing proliferations as the perennial classification of finite simple groups and the likelihood that over a quarter-million theorems are solved each year.[2] But there it is: unwavering and real. The consolidations, perpetuations, and condolences of our predicament offer no solace. We have to face it alone. As the Swiss naturalist Johann Georg Ritter von Zimmermann (1728–1795) contemplated in his (ironically) best-selling work of 1756, *Über die Einsamkeit* (*Of Solitude*), the hermit has the ultimate advantage in life as "Solitude, indeed, inspires the mind with notions too refined and exalted for the level of common life."[3] Zimmermann was not entirely solitudal (any more than was Henry David Thoreau). In 1768 he became George III's private doctor, settling in Hanover on the occasion. The population of Hanover by 1755 was over 17,400 people, hardly a monastery.

Vexatious Variables

In considering a near endless multitude of environmental variables – past, present, and future circumstances – one is most likely disposed toward the mathematical semantics that are wrapped inexorably around some kind of formula. Formulas provide a certain comfort in clarity and the working through of basic mechanical engineering. Yet, all certainty is as elusive as the conditions defining what is known as Kolmogorov's zero-one law within the realm of ergodic probability landscapes. It's a rather grandiose way of picturing the following: The likelihood that something will happen is either zero or one.[4] Ergodic probability relates to what is known as phase space and essentially describes where and when something is happening or will happen, and in what kind of geography, and at which momentum.

Hence, two fundamental situations. There is the one – our theoretical individual – for whom these contemplations toll, amid certainties and uncertainties, probable and improbable futures, alike. Or there is none, a biological nullity, whose statistical likelihood would suggest that evolution herself is no longer relevant, all natural selection one enormous non sequitur, even the barest laws of ecology lacking any lasting traction.

One important mathematical representation of these biospheric conditionals and materializations leads metaphorically from the 5-year-old Russian genius Andrey Kolmogorov (1903–1987) who, among so many of his discoveries (including the above-referenced zero-one law), worked out the object of random variables at the heart of stochastic targets, to the German inventor of set theory and transfinite numbers and his celebrated "infinity of infinities," Georg Cantor (1845–1918). Such mind-bending views of numbers engendered in Bertrand Russell's famed paradox,

---

[2] Hoffman P. The Man Who Loved Only Numbers: The Story of Paul Erdős and the Search for Mathematical Truth. Hyperion, New York, p 204

[3] Published in English by E. Duyckinck, New York, 1819, p 55

[4] See "A Simple Introduction to Ergodic Theory," by Karma Dajani and Sjoerd Dirksin December 18, 2008, http://www.staff.science.uu.nl/~kraai101/lecturenotes2009.pdf. Accessed 23 April 2017

the most standard definition of which declares that it refers to "the set of all sets that are not members of themselves. Such a set appears to be a member of itself if and only if it is not a member of itself. Hence the paradox."[5] Try to place yourself in that formula. It calls upon a strange and almost nauseating insight into what it means to be connected to a species and, by turns, to the Earth.

Equally fascinating (and troubling to many) is our prolific dilemma. Not just what will survive the Anthropocene epoch, but whether within that set of taxons doomed by our one species, we can confidently speak of a *who*. To speak of whom is to assert by implication a relationship. A birth, a death, another birth; a series of individuals connected somehow within an integrally biological configuration: an ecosystem, a community of communities. Mathematical relationships may have ideal characteristics, admirable from afar, like the rings, ringlets, moons, and gaps surrounding Saturn. But indisputably, logic dictates our biological affinities here on Earth, in any place with air and freshwater, modest shelter and at least meager foodstuffs, nothing more. These few bounties are also strict boundaries and herald the existence about which we are and are continuing to reflect upon. There can be little doubting of their veracity: We have a foothold in what is incontestably a *who* and a mathematical and ecological series of who's. That is both who we are and what we do. Our identity ties us to this planet with an ecstatic wonder both empirically given and frustratingly short-lived.

If by "who" we are to incorporate those decisive elements of quantitative genetics, such as the gamodeme (isolated breeding community),[6] quantitative trait loci, and outward phenotypes, then we must with even more rigor consider the impression of environmental variables upon the human heart and its unpredictable pronouncements and initiatives. Our every choice hinges upon such impressionable meta-statistics, wayward passion, and inklings of outcome. By that inward culture (the impressionable components) of intuitive preselection, we perpetually attempt to get beyond our perennial pessimisms, limning our horizons toward greener pasture.

We want to be in control and we are not. No individual, however strict or attenuated its definition, has enough freedom to control the modus operandi of his/her/its circumstances, involvements with other biological organisms, second-by-second nervous system. There is no absolute freedom anywhere in the universe, if the laws of physics across 13.8 billion light years since the Big Bang are to be believed. We are the focus and full footprint of our own dream of freedom, fully aware that we have ten fingers, not eleven (as a rule); a range of other anatomical and physiological parameters that are unwavering, short of some sudden upset, which, again, we had not predicted. Freedom beyond such freedom is an illusion which we are free to imagine, but are refused entry into its actuality because it has ceased to exist the instant we picture it, other than in actual picture books of one form or another.

---

[5] Stanford Encyclopedia of Philosophy. https://plato.stanford.edu/entries/russell-paradox/. Accessed 23 April 2017

[6] See Mann DG (1989) The species concept in diatoms: evidence for morphologically distinct, sympatric gamodemes in four epipelic species. Plant Syst Evol 164:215. https://doi.org/10.1007/BF00940439, https://link.springer.com/article/10.1007/BF00940439. Accessed 23 April 2017

Fantasy. The history of art. That picture – Utopia – has become an object within our sense of freedom. Once confined to a sense, it is without any more substance than any other idea. However fixated we may be with our apotheosis of its emblems and incarnations, these countless iterations do not alter the truth of our subjugation to our own Self. We cannot moderate a conversation between a real Utopia and our *idea* of Utopia, not unlike the paradox of the Greek city of Megalopoli in Greece's southwestern regional Arcadia, in Peloponnese. It is no pastoral paradise today, as the Renaissance thought of her, with its nearly 5800 people and a freeway by which one can reach Athens in under 2 h. Moreover, the city's outskirts of Valtetsi contain Greece's second largest lignite mine, where high levels of radioactive ash in the air have been a persistent problem, first officially acknowledged in a Greek national study in 2002.

Such ironic roots remind us that we are an individual of inventive proclivities, but also the same individual who loses out by the very distance or duration we have engendered and now must endure, as someone isolated from that which we had hoped to inhabit. Ecology, that interdisciplinary science of human alienation places yet another clouded window on the aforementioned Russell's Paradox. Its cruelty grows as we increasingly vanish as Selves. Every decision we enact can be characterized as mathematically moving us farther away from the source of our original being. There is nothing we can do about it, short of precocious, collective choices.

"Readiness Potential"

It has been suggested that people "make 773,618 decisions over a lifetime – and will come to regret 143,262 of them. A typical adult makes 27 judgments a day – usually starting with whether to turn off the alarm or hit snooze."[7] That would be based upon a lifespan of approximately 80 years. At a rate of approximately 20% regret, does that decision ratio influence or accelerate positive choice predilections? We cannot lend much certainty to this discussion because we have no data on the distribution of regrets. Is the 20% of 80 years (16 years) continuous? Probably not. Intermittent, to be sure. Weighted in youth, middle age or among the elderly?

All such statistics reveal great variance. For example, "Various internet sources estimate that an adult makes about 35,000 remotely conscious decisions each day (in contrast a child makes about 3,000). This number may sound absurd, but in fact, we make 226.7 decisions each day on just food alone...."[8] But herein we hit up against semantics: Are choices the same thing as decisions? And what about "remotely conscious"? Again, as a reminder, we choose this course of discussion

---

[7] By Mirror.CO.UK, 11 NOV 2011, http://www.mirror.co.uk/news/uk-news/average-person-makes-773618-decisions-90742. Accessed 22 April 2016

[8] "35,000 Decisions: The Great Choices of Strategic Leaders," posted by Dr. Joel Hoomans on Mar 20, 2015, Roberts Wesleyan University, http://go.roberts.edu/leadingedge/author/dr-joel-hoomans. Accessed 22 April 2017

given the premise that our very individuality hinges upon our idea of freedom (as elegantly detailed by Sartre and others) and that freedom is invariably tied to choices, whether our own choices or choices by others – other individuals, other forces (e.g., laws, regulations, cultural customs, bio-heritage, peer pressure, causes and consequences of desire, and attachment).

With so much deliberation and uncertainty affecting individuals of our species, it would seem likely that outcome-orientations are rooted in a combination of intuitive and cognitive conjuration. In 2005 a team of researchers led by Carlos Bustamante of the Department of Biological Statistics and Computational Biology at Cornell University conducted a massive study of DNA from 11,000 genes among 39 people, hoping to better understand positive and negative (weak) influences on the mechanisms of natural selection that drive molecular evolution within a given species. Their findings suggested that 9% of all human genes were rapidly evolving: "304 (9.0%) out of 3,377 potentially informative loci" revealed compelling proof of "rapid amino acid evolution."[9] If nearly one-tenth of all genetic materials within our species are vulnerable to rapid change, then we may assume that choices and the illusion and/or reality of freedom teases into being something like a biological Russian roulette. It means that, at the very least, an individual, so-called, has a one-in-ten chance of changing a significant personal paradigm within his/her lifetime or of passing down to another individual, an offspring, the propensity for such a change. It also connotes a very real possibility that the evolution of amino acids, 20 out of 500 or so fixed in any human being's genetic code, respond by some, as yet completely unknown chain of events (building blocks of change), to the choices a person makes during his/her lifetime.

Or, more complex still, that the sum total of choices preceding a choice at any given instant might pass the threshold for transformation, much like that turning point whereby the mass of a particle of stellar dust falling to Earth can actually be measured and weighed, the number varying from "10–16 kg" to "10–4 kg."[10] It has also been estimated to weigh "0.000,000,000,753 kg."[11]

Similar to the dust particle is a grain of sand. That grain has been measured by NASA. If mass is the measure of density multiplied by volume, then one grain of sand equals approximately 1.1 x10–13 kg/grain.[12] The same grain, according to

---

[9] See Letter, Nature 437, 1153–1157 (20 October 2005) | https://doi.org/10.1038/nature04240; Received 24 April 2005; Accepted 14 September 2005, "Natural selection on protein-coding genes in the human genome," Carlos D. Bustamante, Adi Fledel-Alon, Scott Williamson, Rasmus Nielsen, Melissa Todd Hubisz, Stephen Glanowski, David M. Tanenbaum, Thomas J. White, John J. Sninsky, Ryan D. Hernandez, Daniel Civello, Mark D. Adams, Michele Cargill & Andrew G. Clark. http://www.nature.com/nature/journal/v437/n7062/abs/nature04240.html. Accessed 22 April 2017

[10] "Spacecraft Measurements of the Cosmic Dust Flux," Herbert A. Zook. https://doi.org/10.1007/978-1-4419-8694-8_5, https://link.springer.com/chapter/10.1007%2F978-1-4419-8694-8_5. Accessed 22 April 2017

[11] "Scientific Notation," https://www.nyu.edu/pages/mathmol/textbook/scinot.html. Accessed 22 April 2017

[12] https://spacemath.gsfc.nasa.gov/Modules/8Mod3Prob1.pdf. Accessed 18 April 2017

classical Jain science, contains a Nigoda, a member of the class of Jiva's atom-sized souls. A Nigoda has a sense of touch. It is alive, an individual that may or may not be part of a cluster.[13] Hence, our interest in NASA and the weighing of astronomically minute particles of dust. The cluster equates with a physical and/or ecological community. One could analogize to the Amazon or Great Barrier Reef. To all of those vast life forms that congregate as individual parts of a whole, from a section of *Armillaria solidipes* honey fungus in Oregon spanning 3.7 square miles[14] to the Aspen grove in Utah comprising "6000 metric tonnes" and "106 acres."[15] All such trivia is more than merely interesting, in our minds: it is intensely relevant to our lives.

The inception of a decision and the chain of events that cumulatively call such relevancies into action are metaphorically rooted to this invisible-to-the-eye dust mote – a Jiva, a grain of sand – that suddenly captures the attention of a specific mass and has energy, can be measured and made whole, like those chunks of icy water mixed with small rocky particles around Saturn.

Inevitably, the mechanics of choice are synonymous with a variety of individual types, each one equally critical to life cycles at an ethical level. We suggest this by way of speculation because it seems likely that ethical chains of events linked expressly to ecological, geo- and astrogeophysical sequences are the only scenarios that weigh more meaningfully for the biosphere than that of the human ego.

Take that in and breathe. It is a massive conundrum, certainly for the ethicist and the metaphysician. But equally so for the legislator, linguist, psychologist, and judge.

Moreover, only at that interface of individual choice and whole ecosystems do nuts and bolts get transformed into the causes and effects of potential convictions. Are epiphany and revelation more efficacious fuels than natural selection and evolution? Are individuals more meaningful than communities of individuals? Or are the reverse better mirrors of reality and blueprints for the future? We don't know. Maybe we can never know. Knowing with certainty or not is less relevant than the pivotal truth of the action itself. By whatever means the mechanism of an action is excited, the topology of its itinerary, its tears and manglings, the many bends and general landscape properties collectively are proof that the liabilities of existence are manifest in any single being. The lone prairie dog emerging from his/her burrow and recognizing a hawk. A ruby throated hummingbird heading out to sea from the coast of Florida toward the Yucatan, a 48-hour arduous epic. It may be an annual pulse hardwired into the little girl's migratory circuitry, but she - that hummingbird weighing, on average, 3 grams - is as alone in this move as in any Kierkegaardian

---

[13] http://www.jainworld.com/book/qaonjainism/ch22.asp. Accessed 18 April 2017

[14] BBC, Earth, "The Largest living thing on earth is a humongus fungus," by Nic Fleming, 19 November 2014, http://www.bbc.com/earth/story/20141114-the-biggest-organism-in-the-world. Accessed 23 April 2017

[15] "The World's Largest Known Organism in Trouble," Living On Earth, February 1, 2013, *PRI's Environmental News Magazine*, http://www.loe.org/shows/segments.html?programID=13-P13-00005&segmentID=7. Accessed 23 April 2017

leap. The steely virtuous baker who begins his day at 4 a.m. or so, as he/she has done all of their life. A family to feed. That this is being, for purposes of description is, indeed, an individual. And that this individual, by dint of the reciprocity potential sustained between individuals (of every species), will necessarily make choices.

The psychologist Benjamin Libet (1916–2007) developed an experiment to determine the time between subconscious buildup of a decision leading to a response that had acted on that decision. His results showed that human subjects needed some "200 milliseconds" between conscious volition and actual action (as recorded by an EEG, a buildup process that has come to be known as the "Bereitschaftspotential" or readiness potential). In fact, that readiness might occur for as long as 7 s prior to the subject even becoming aware of the decision he/she had made.[16] In concert with readiness potential, Tübingen neuroscientists, Anna-Antonia Pape, and research group leader Markus Siegel of the Werner Reichardt Center for Integrative Neuroscience (CIN) and MEG [magnetoencephalography] Center have "found a neural correlate… in the motor cortex itself. They showed that the upcoming motor decision can be predicted from the status of motor areas even before decision formation has begun."[17]

By that indicator, we might also arrive at that juncture where we can predict choices, calculate ethical considerations and even their outcomes. Should the sciences suggest this level of certainty, measured in milliseconds and seconds, does it change anything fundamentally for us? Would that make free will a thing of the past? Again, we don't know. The data might abet invasions of privacy, such as was portrayed in the fictional crime prevention scenario laid out in Steven Spielberg's 2002 film, "Minority Report." Polls, demographic and health surveys, elections, branding, and advertising in general might well lay claim to a whole new science following upon certainties previously unfathomed. With readiness potential in any individual given, would that carry over to a similar readiness potential in whole communities? Nation states? If the day is near that empires can be predicted, then they can also be manipulated, but by who, by what? And what motives will be at play?

The question of humanity comprising individuals leads through a labyrinth of unknown and heavy morasses. Philosophy cannot hope to be fully equipped to grap-

---

[16] See Keim, Brandon (April 13, 2008). "Brain Scanners Can See Your Decisions Before You Make Them." Wired News. CondéNet and Chun Siong Soon, Marcel Brass, Hans-Jochen Heinze, John-Dylan Haynes, April 13, 2008. In addition, "Unconscious determinants of free decisions in the human brain (Abstract)." Nature Neuroscience. Nature Publishing Group. 11(5):543–545. https://doi.org/10.1038/nn.2112. PMID 18408715. See also, Wegner D., 2002. The Illusion of Conscious Will. Cambridge, MA: MIT Press

[17] Tübingen University "Do We Really Have A Choice?" NeuroscienceNews. NeuroscienceNews, 7 October 2016. http://neurosciencenews.com/neuroscience-choice-psychology-5239/. Accessed April 17, 2017. See also "Mindfulness Can Help You Avoid Self-Destructive Decision-Making," cited in "The Neuroscience of Making a Decision – Various brain regions work together during the decision-making process." Posted May 06, 2015, by Christopher Bergland, https://www.psychologytoday.com/blog/the-athletes-way/201505/the-neuroscience-making-decision. Accessed 17 April 2017

ple with all the loose ends and, possibly, central pillars of a new faith that relies on certainty as opposed to uncertainty; the tyrannies of manipulation versus free will; and that mortifying authority which might undermine – as had happened so frequently throughout history – any or all individual choice.

A Psychiatric Black Hole

In their study of decision-making, Monique Ernsta and Martin Paulus examined three stages that they have deemed critical: "(1) formation of preferences among options, (2) selection and execution of an action, and (3) experience or evaluation of an outcome."[18] While this map seems overtly logical, there is, in fact, no predictable end to speculation regarding contingent realities bearing weighty outcomes.[19] That is because the embodiment of individualism within the entity of a philosophical singularity, a body, a human being, any singular organism that has managed to separate or individuate from the community – a single coral polyp, for example, or a leaf – all invite consideration of the noumenal, or das Ding an sich, a "thing in itself," as Immanuel Kant characterized that earth-bound truth, good for at least another four billion years of the planet's potential.

Choice, in this respect, is that very thing in itself, not a reaction time, or reactive potential, but a transitive, active ontology that may be a lifelong nurturance or instantaneity. The point is, it (choice) may enshrine every nuance associated with freedom.

Freedom comports with both a principle and an individuated object, as qualified in Edward Zalta's groundbreaking work, *Abstract Objects: An Introduction to Axiomatic Metaphysics*,[20] one of the only actual "theories" in all of philosophy, as versus the many hundreds of theories in mathematics, physics, complex analysis, topology, group theory, probability, statistics, and differential geometry. The ensuing linguistic, mathematical, and conceptual debate assisting "the identity-conditions of properties"[21] has been tested by linguist Dirk Greimann against logician/philosopher Willard Van Orman Quine's famed proposition, "No entity without identity" [that corresponds with] "the ontological recognition of a sort of objects f (sets, properties etc.)" [which] "is legitimate only when the fs have been provided with a sat-

---

[18] Science Direct, Ernsta M, Paulus MP (2005) Neurobiology of decision making: a selective review from a neurocognitive and clinical perspective. Biol Psychiatry Rev 58(8):597–604. http://www.sciencedirect.com/science/article/pii/S0006322305007109. Accessed 17 April 2017

[19] See, for example, http://whatculture.com/history/10-seemingly-insignificant-choices-changed-world-forever; and http://www.cracked.com/article_18644_5-world-changing-decisions-made-ridiculous-reasons.html

[20] Abstract objects: an introduction to axiomatic metaphysics. D. Reidel Publishing Company, Dordrecht/Holland, 1983

[21] Greimann D (2003) Is Zalta's individuation of intensional entities circular? Metaphysica 2(2):93–101. https://philpapers.org/rec/GREIZI. Accessed 6 May 2017

isfactory individuation."[22] Zalta's enormous contribution to this arcane but crucial debate, the quintessential understanding of individuation,[23] hinges upon "worlds."[24] Subsets of these worlds include "non-normal worlds," "nonclassical worlds," and "non-standard worlds."[25]

Quine (1908–2000) brilliantly expanded this identity problem by asserting, "To be is to be the value of a variable."[26] It was in Quine's book *Word and Object*[27] that the great analytical philosopher introduced the notions of the "indeterminacy of translation" and the "inscrutability of reference," thereby challenging all sets, subsets (f's), and all forms of empiricism as propagated within linguistic relations (e.g., language and representation, sentences and referentials, and the Sorites Paradox – first-order predicate calculus and logic, the famed heap of sand being the most obvious reference delusion. At what point does a heap of sand, having blown to virtual nothingness – say one grain of sand left – go from being a "heap" to an individual?). Quine considered philosophy a form of science, which, for our purposes, has great merit because of his remarkable suggestion that all translation and reference is powered by a distinct "indeterminancy."[28]

Gregory Rabassa, who effected a remarkable translation of García Márquez's *One Hundred Years of Solitude*[29] spoke of the translator as one afflicted with "induced schizophrenia." We should feel no less inclined to radical shifts in openly opining upon that gulf harboring both Self and Species, Individual and Community. All are theorems whose solutions are vested in one bias or another. Writes, Joye Weisel-Barth, "We live in an era of the vanishing self."[30]

To choose or not to choose would at once go toward some level of confirming adaptive responses. Research into this region of logic, physiology, and neurobiology reveals the salient role of various parts of the brain, dark holes where no psychiatrist or psychologist or philosophy has more than a fleeting chance of penetrating. If the brain indeed contains locations specialized in aggregate in the formation of all those

---

[22] http://www.metaphysica.de/texte/mp2003_2-Greimann.pdf. Accessed 18 April 2017

[23] See *Principia Logico-Metaphysica* (Draft/Excerpt) Edward N. Zalta Center for the Study of Language and Information Stanford University October 28, 2016 http://mally.stanford.edu/principia.pdf. Accessed 18 April 2017

[24] 12.4, "Impossible Worlds," https://mally.stanford.edu/principia.pdf. Accessed 18 April 2017

[25] Ibid., *Principia Logico-Metaphysica*

[26] Selection from Milton Munitz (1981) *Contemporary Analytic Philosophy*: "On What There Is," Macmillan Publishers, London. http://www.loyno.edu/~folse/Quinewhatis.html. Accessed 18 April 2017

[27] MIT Press, Cambridge, Mass., 1960

[28] See his book *Ontological Relativity and Other Essays*, Chapter 2: Ontological relativity," pp 26–68, Columbia University Press, New York, 1969

[29] Lucas Rivera. "The Translator in His Labyrinth." Fine Books Magazine, https://www.finebooksmagazine.com/issue/0104/translator.phtml. Accessed 7 May 2017

[30] "Stuck: Choice and Agency in Psychoanalysis," Joye Weisel-Barth Ph.D. and Psy.D., Pages 288–312 | Published online: 07 Jul 2009; https://doi.org/10.1080/15551020902995306, http://www.tandfonline.com/doi/abs/10.1080/15551020902995306?journalCode=hpsp20. Accessed 19 April 2017

complexities that make for decisions, judgments, and most of all, choices, the gravity of a philosophical injunction is no less for it. Nor are the ethical implications, regardless of whether the anatomy decrees a millimeter to the left or right, such as in the orbitofrontal cortex in the human brain,[31] or those countless "neurons in the parietal cortex [which] reveal the addition and subtraction of probabilistic quantities that underlie decision-making,"[32] Added to which are revelations suggesting that the "left dorsolateral prefrontal cortex (DLPFC)" has been implicated in human decision-making.[33]

Without specifying precise anatomical loci, other researchers have suggested the mere presence of the subconscious (which might as easily mean Carl Jung's "collective unconsciousness," or Mircea Eliade's archetypes and repetitive motifs throughout human anthropological history) as those generalized portions of thought attributable to both the action and function of choice in the human psyche and, potentially, human evolution.[34]

The physiology that ultimately may decree choice, and its miraculous mechanisms, present an edifice of endless fascination. But whether our decisions can one day be deduced by precise locational analysis, or a comprehensive understanding of the engineering wired into our overt decision-making, the research will always lack for the 365 degrees which necessarily argue for the individual.

In 2002 the Nobel Prize in Economics was awarded to Daniel Kahneman (his close colleague, Amos Tversky, had passed away by then but would have shared in the prize). Their ironic research regarding so-called prospect theory involved human choices that encompass editing, elimination of bias (framing), and evaluation of risk; proposing that in spite of any and all prospect trawling by the human brain, our species is not good at rational decision-making. Contrary to this gospel of human cognitive flaws, an associate professor of brain and cognitive sciences at the University of Rochester, Alex Pouget, has shown that "people do indeed make optimal decisions – but only when their unconscious brain makes the choice." The unconscious brain turns out to make probability computations with greater ease and accuracy than the conscious mind; but those computations are absorbed by

---

[31] See Takahashi YK, Roesch MR, Stalnaker TA, Haney RZ, Calu DJ, Taylor AR, Burke KA, Schoenbaum G (2009) The orbitofrontal cortex and ventral tegmental area are necessary for learning from unexpected outcomes. Neuron 62(2):269–280. https://doi.org/10.1016/j.neuron.2009.03.005, https://www.ncbi.nlm.nih.gov/pubmed/19409271. Accessed 19 April 2017

[32] Yang T Shadlen MN (2007) Probabilistic reasoning by neurons. Nature 447(7148):1075–80. Epub 2007 Jun 3. https://www.ncbi.nlm.nih.gov/pubmed/17546027, PMID: 17546027. https://doi.org/10.1038/nature05852. Accessed 19 April 2017

[33] See Rorie AE, Newsome WT (2009) A general mechanism for decision-making in the human brain?" Trends Cogn Sci 9(2):41–43. Rorie AE, Newsome WT (2005) A general mechanism for decision-making in the human brain? Trends Cogn Sci 9(2):41–43. PMID: 15668095, https://doi.org/10.1016/j.tics.2004.12.007, [Indexed for MEDLINE], https://www.ncbi.nlm.nih.gov/pubmed/15668095. Accessed 19 April 2017

[34] "Our Unconscious Brain Makes The Best Decisions Possible," December 26, 2008, University of Rochester, http://www.rochester.edu/news/show.php?id=3295. Accessed 19 April 2017

consciousness at the level of a "triggering threshold" and tend to be given greater weight in decision-making.[35]

That threshold has undergone study in the form of perceived choice mechanisms in primates other than human. "By understanding how other species' choices go awry, we may be better able to both understand and improve the decisions of our own species."[36]

Such mechanisms have also been studied in migratory birds. Theunis Piersma, Professor of Global Flyway Ecology at the University of Groningen, states, "It could well be that there is a relationship between the moment a bird was born and the moment it wants to breed. Distinguishing such environmental factors from genetic influences is far from easy. To find out the extent to which the environment and life experiences played a role in the choices that a bird makes requires you to follow them from cradle to grave."[37]

The same "triggering threshold" had been applied to algorithms that incorporate so-called small error, analysis that approximates with the statistical underpinnings of regression analysis, the grasping of relationships among variables, with a clear line drawn in the sand between numerical stability which refers to "well-conditioned problems" versus "ill-conditioned" ones. "An algorithm is called numerically stable if an error, whatever its cause, does not grow to be much larger during the calculation. This happens if the problem is *well-conditioned*, meaning that the solution changes by only a small amount if the problem data are changed by a small amount. To the contrary, if a problem is ill-conditioned, then any small error in the data will grow to be a large error."[38] The Authors write, "we have argued that ineliminable noise in neural systems requires the agent to make certain kinds of commitments in order to make decisions, and these commitments can be thought of as the establishment of policies. Noise puts a limit on an agent's capacities and control, but invites the agent to compensate for these limitations by high-level decisions or policies that may be (a) consciously accessible, (b) voluntarily malleable, and (c) indicative of character. Any or all these elements may play a role in moral assessment."

We have thereby moved from the inception of choice to its cultural component, namely, the virtuous act. In every Golden Rule and book devoted to such hallmarks of so-called civilized behavior, the entablatures of morality are contested at every junction by a Naram-Sin (an ancient Mesopotamian leader who is seen in an Iraqi

---

[35] See http://www.rochester.edu/news/show.php?id=3295, PR3295, MS 1775. Accessed 19 April 2017

[36] See Annu Rev Psychol. Author manuscript; available in PMC 2015 Jun 2. Published in final edited form as: Annu Rev Psychol. 2015 Jan 3; 66: 321–347. https://doi.org/10.1146/annurev-psych-010814-015310, PMCID: PMC4451179 NIHMSID: NIHMS692900 "The Evolutionary Roots of Human Decision Making," Laurie R. Santos and Alexandra G. Rosati, https://www.ncbi.nlm.nih.gov/pmc/articles/PMC4451179/. Accessed 19 April 2017

[37] See "How migratory birds make choices," https://www.nwo.nl/en/research-and-results/cases/how-migratory-birds-make-choices.html. Accessed 19 April 2017

[38] See Neurosci., 23 April 2012 | https://doi.org/10.3389/fnins.2012.00056, "The neurobiology of decision-making and responsibility: Reconciling mechanism and mindedness," by Michael N. Shadlen and Adina L. Roskies Frontiers in Neuroscience, http://journal.frontiersin.org/article/10.3389/fnins.2012.00056/full. Accessed 17 April 2017

stele attempting to conquer with his spear and his army a whole mountain) or any other despot whose ignoble motivations or heinous deeds history has seen fit to record.

Says C. Daniel Salzman, "We really don't understand at a systemic level how psychiatric disorders result from dysfunction of neural circuits that produce different cognitive and emotional symptoms, and that thereby affect decision making. We also don't understand how psychiatric treatments work -- how they change neural activity in the brain. The reason we know almost nothing about this is because we really don't know in detail enough about how those neural circuits work normally."[39]

If we remain at a loss to understand the holocaust, what can we ever know about the countering mechanisms in an Oscar Schindler or Mahatma Gandhi? Researchers are scrambling to discern any hierarchy of certitudes, all those vague mechanisms in the brain that determine choice. But what kind of choices, if not those strictly oriented in the Darwinian and Mendelian ethos? As long as we are *Homo sapiens*, is there any chance that self-consideration can achieve clarity, despite the accumulation of inevitable and important biases? After thousands of years of reflection, we are still lodged somewhere in the psychiatric black hole of determinations that are inside the subject, not outside. There is no objectivity and no clue as to what is really behind the human penchant making a wish, or declaring that "It's my turn to choose," or obstinately clinging to such demands as "I want that one." Or choosing forgiveness rather than revenge.

We simply do not understand, ultimately, what compels a person to make a decision. There are key impulses that are instantaneous, to be sure. Studies have shown that even severely afflicted schizophrenics, suffering from a hypervariance of maladies culminating in their state of being nonetheless, will likely flee in the event of a fire, like almost any other individual. When we are out hiking in a wilderness and a sound of heavy rushing through the underbrush comes at us suddenly, one (not both) of our hands react with a near instant inward clutching, a fist.

Far more significant but less clear are the fundaments of discernment and delineation that quickly move us to unclench the fist; the roots of compunction, softness, kindness, and mercy; or the etiology of the poet Robert Frost's famed, "Two roads diverged in a wood and I - I took the one less traveled by, and that has made all the difference."[40]

---

[39] "The Neuroscience of Decision Making" The Kavli Foundation, http://www.kavlifoundation.org/science-spotlights/neuroscience-of-decision-making#.WPWzrlPyvw4, Accessed 17 April 2017

[40] See https://www.brainyquote.com/quotes/quotes/r/robertfros101324.html, Accessed May 10, 2017. See also Frost's collection, *Mountain Interval*, 1916; see The Paris Review, September 11, 2015. See also Robinson, Katherine. "Robert Frost: 'The Road Not Taken'." Poetry Foundation. Poetry Foundation

Judgment

For purposes of conceding any comprehensive understanding while managing to rummage through the day with as benign and compassionate a perspective and personal footprint as possible, let us begin by suggesting a relatively simplistic yet relevant insight. The only antidote to ignorance and violence in whatever forms that seem to attack us converts to a near mathematical surety: each victim must be recognized. Recognition cuts across all levels of communication and translation. We have remarked upon the "indeterminancy" of translations, but that is an obtuse poetic, analytically speaking. In reality, in any reality, recognition entails the oldest adages advanced in favor of treating others as you would have them treat you. A community of individuals who know by intuition, by logic, by experience, by common sense, that the bulwark of their happiness is in both the recognition of others and with it the kindness that comes from moral judgment. The judgment is not about judging so much as it is the *choice to be generous*. Jurors and judges have their day, and for those mired in cold calculus, the whole world is easily transmogrified into a courtroom. But we also agree to agree at every stop sign, at each instance of "yes" and of "no." We are each of us complicit by dint of our biological standing in relationships that recognize us, whether we like it or not. This fact alone deciphers the stamp of the individual in group mind theory, as well as at the polling booths. It pulls from the trenches a wounded soldier. Leaps into floodwaters to rescue the three cows marooned on a floating raft in New Zealand several years ago. Or pulls the dog from a burning building.

We have ample archetypes to inform whole universities of learning with iconic memories of heroism in daily life and – with every cumulative crisis – more and more opportunities to let our individual voices be heard. The fracas of vulnerabilities calls out to us for collective judgment: to ban nuclear weapons and the killing of other animals. To cease the destruction of the biosphere through our callous indifference and greed. These injunctions are unambiguous.

What is much less clear is how the individual – an Edward Curtis, for example, as discussed earlier – comes to that revelatory moment empowering him to do what is both visionary but also what is right. From the countless annals of anthrozoology, we know with certainty that nurturance is not simply a word instinct within the mammary glands but an active chain of events inherent to the brain (in whatever form or location) and to all life, whether we think of specific individual instances of parasitism or co-symbiosis. From our severely limited vantages as a species, variations in the genome and metamorphoses of all embryology are full of surprises. As when a butterfly takes flight or a flying fish for that matter. Evolution has provided a fitting metaphor for any number of interpretive possibilities as individuals and communities grow up. But no theory can refute a clear piece of advice: Evolution does not condemn us. Only our choices can do that. But in acknowledging this more than likely fundament of human behavior we must also recognize that "One prospect cannot occupy two different positions." This profound suspicion was voiced by John Broome in a subchapter entitled "Individuating Outcomes," in an anthology titled *The Limits of Rationality*.[41]

---

[41] *The Limits of Rationality*, edited by Karen Schweers Cook and Margaret Levi, John Broome's

Those choices, and the judgment that has invested a lasting hope in such decision-making, are the sole salutations in our arsenal of survival and dignity.

There may, indeed, be some theoretical point of view that gives us pause when we consider the future individual whose role in the biochemical collective means everything to everyone or at least someone. That theoretical individual lies ready and waiting in each of us. Peter Marris has written of places and people we love, "These specific relationships, which we experience as unique and irreplaceable, seem to embody most crucially the meaning of our lives."[42] It remains a point of scientific inscrutability that solitude, attachments, and meaning that underscores the reality of most sentient lives, including our own, remain crucially tied to and were born of macromolecules; by all of those infinite other connections that do not, on the surface, seem to have any relevance to our daily goings-on. Such biochemical events as light absorption, electromagnetic waves, the primordial coordination of nucleotides, the "template polymerization of biopolymers,"[43] each action of the universe informing everything we do, all make us who we are and prepare the groundwork for what we might choose to do. And just as these physical, unseen civilizations of physics, of light, space, matter, and energy are not those we could count as relatives or even members of our class of organisms, we know well enough that there are laws of physics worth respecting – like not whimsically, or in an ill-considered manner, leaping off the summit of El Capitan in Yosemite National Park. But there are equally other nonfamily member laws of life that are as daunting and salient within our ephemeral moments here on Earth. Writes John Cohen in his groundbreaking work on the mind, "When we imagine, we are aiming at an object which is non-existent, or, at least, absent; we act in this way by means of its physical or mental content which is present only as a representative of the object."[44] Similarly, when we ask of the individual to uphold his/her most compelling argument for being, namely, the conscience, which has been perpetually proposed as a philosophical possibility rather than some biological component or necessity, we are also presuming a major leap of consciousness. As the most interesting politician/philosopher, T. V. Smith, wrote in his compelling *Beyond Conscience*, "If conscience can invent the group which validates its dicta, then we have really changed the venue of our validation from sociality to prophecy...."[45] Helping to galvanize a seer's energies, imagination and freedom is not just the singular duty of ecological

---

essay contribution, "Should a Rational Agent Maximize Expected Utility?" The University of Chicago Press, Chicago, Ill., 1990, p 138

[42] Peter Marris, "Attachment and Society," in *The Place of Attachment in Human Behavior*, edited by C. Murray Parkes and J. Stevenson-Hinde, London, 1982, p 185. Cited in *Solitude – A Return to the Self*, by Anthony Storr, Balantine Books, New York, 1988, p 12

[43] *Information in biological systems: the role of macromolecules*, by Werner Holzmüller, translated by Manfred Hecker, Cambridge University Press, Cambridge, UK, 1984, p 25

[44] See John Cohen's *The Lineaments of Mind in Historical Perspective*, W. H. Freeman and Company, Oxford and San Francisco, 1980, p 239. Cohen is, in part, reflecting upon Jean-Paul Sartre's work, *The Psychology of the Imagination*, Methuen, London, 1972, p 20

[45] Smith TV (1934) Beyond Conscience. Whittlesey House/McGraw-Hill Book Company, Inc., New York/London, p 137

citizenry, but our only hope of survival. For every one of us. Us. Individua*ls*. The proof of such far-reaching speculations comes down to a very basic informational unit that affects us in countless ways, every day. Robert E. Goodin articulated this back-to-basics quantum as follows: "Uncertainty may be a plausible excuse for individuals, but not for collectivities. With the pooling of facts and resources, the group naturally has both more information and more resources for enhancing information than do any of its members. Besides, what is uncertain from one perspective is often more certain from another. For example, we can know that one thousand people will die this year in traffic accidents without knowing who the victims will be."[46]

Not Who but What?

With the group mind having bolstered our illusion of freedom over a previous duration of not just 700 million but several billion years, we have arrived at a neurological commons that is not the *who* – someone standing alone in a Times Square – but the *what*: What is it about any probable cause of an individual, in both the legal identity, genomic, ethical, and evolutionary sense, that ties up so many years and lifelines into a bundle of inexplicable potential? The question falls apart, as the poet Yeats likened the syndrome, in the sense that the actual life span of an individual, even the most ancient of life forms, is biochemically trivial in the context of astrogeophysical numbers. No individual "who" is ever going to transcend the most critical part of it: the "what." What was it? What was he/she all about? We remember the faces of Buddha and Christ only because of the multitudinous extent of their depictions, up unto the present time (in human time). But imagine the subtlest, grandest, bountiful and beauteous, sovereign and elegant, the most remarkable of bacteria, of dinosaurs, and of plants. Microscopic, enormous, modestly grown, three individuals – bacterium, dinosaur, plant – whose names, species, time frames are now utterly unknown and/or irrelevant to our thinking, feeling, and daily intentions and decisions, even though it is likely that we exist because those three anonymous ones, teeming with cellular excitement and genetic restlessness, existed. And many more like them and unlike them.

So that by the "what" we suggest, of course, the sum total of choices and that community of like-hearted predilections – tied to members of the only biosphere we know of, at present – whose origins and present tense behavior all suggest some kindly organism, of organisms that behave decently; that act resiliently and, as the situation doth demand, selflessly, out of *true love*, a wondrous phrase mightily overused such that we might be tempted (for academic permutational purposes, so to speak) to hazard some other linguistic or conceptual embodiment in so confusing and magnificent a world. An entirely different paradigm of original characterization. But, to be clear, there isn't one.

---

[46] Goodin RE (1985) Protecting the vulnerable: a reanalysis of our social responsibilities. The University of Chicago Press, Chicago/London, p 138

The manufacturer's authorised representative in the EU is Springer Nature Customer Service Centre GmbH, Europaplatz 3, 69115 Heidelberg, Germany. If you have any concerns regarding our products, please contact ProductSafety@springernature.com

Printed and bound by CPI Group (UK) Ltd, Croydon, CR0 4YY

23/03/2026

02076443-0005